Trees in England

Trees in England
Management and disease
since 1600

TOM WILLIAMSON,
GERRY BARNES
AND TOBY PILLATT

UNIVERSITY OF HERTFORDSHIRE PRESS

First published in Great Britain in 2017 by
University of Hertfordshire Press
College Lane
Hatfield
Hertfordshire
AL10 9AB

British Library Cataloguing in Publication Data
A catalogue record for this book is available from the British Library

ISBN 978-1-909291-96-6

Design by Arthouse Publishing Solutions Ltd
Printed in Great Britain by Charlesworth Press, Wakefield

Contents

Abbreviations

Barn	Barnsley Record Office
Bev	Beverley Record Office
Brad	Bradford Record Office
Bury St Edmunds RO	Bury St Edmunds Record Office
Cald	Calderdale Record Office
Don	Doncaster Record Office
DRO	Dorest Record Office
ERO	Essex Record Office
HALS	Hertfordshire Archives and Local History
HAR	Herefordshire Archives and Records
Hull	Hull Record Office
Ipswich RO	Ipswich Record Office
KHL	Kent History and Library Centre, Maidstone
Kirk	Kirklees Record Office, Huddersfield
Lds	Leeds Record Office
NHRO	Northamptonshire Record Office
NRO	Norfolk Record Office
Nrth	Northallerton Record Office
PRO/TNA	Public Record Office/The National Archives, Kew
SA	Shropshire Archives
Shf	Sheffield Record Office

Figures

Acknowledgements

A large number of individuals and organisations have helped with this book. We would like to thank, in particular, the Arts and Humanities Research Council, DEFRA and the Woodland Trust, who provided financial support for much of the research presented here, and the University of East Anglia, who provided support towards publication costs. We are also grateful to Nick Atkinson, Peter Austin, Karen Barnes, Gary Battell, Liz Bellamy, Teresa Betterton, Richard Brooke, Peter Clarke, Sid Cooper, Patsy Dallas, Paul Dolman, Andrew Falcon, David Fox, Rob Fuller, Justin Gilbert, Jon Gregory, Rory Hart, Robert Liddiard, Jim Lyon, Andrew MacNair, Sam Neal, Nicola Orchard, Crispin Powell, Mark Pritchard, Rachel Riley, Ian Rotherham, Anne Rowe, Sarah Rutherford, Steve Scott and Adam Stone, all of whom provided help, advice, encouragement or information. Lastly, we would like to thank the staff of Essex Record Office, the Herefordshire Archives and Records Centre, Hertfordshire Archives and Local History, Kent History and Library Centre, Norfolk Record Office, Northamptonshire Record Office, Shropshire Archives, the Suffolk Record Offices at Ipswich and Bury St Edmunds, the National Archives and the various Yorkshire record offices for all their help and assistance.

The photographs, maps and diagrams are our own, except: 1.5 (geog photos/Alamay Ltd); 2.5 (National Trust); 1.3, 7.5 and 8.3 (Steve Scott/Forestry Commission); 3.4 (Vivienne Blakey); 4.3, 4.9, 5.8 and 6.5 (Ian Rotherham); 4.5 (Sarah Rutherford); 5.2 (David Hosking); 6.7 (Jon Finch); and 7.7 (Richard Revel).

Figures 1.10, 2.1, 3.1, 4.2 and 4.7 are reproduced with the permission of Norfolk Record Office; 5.1, with that of Essex Record Office; 2.3 is courtesy of Northamptonshire Record Office; 2.4, 5.7 and 8.5 are courtesy of Hertfordshire Archives and Local History. Figure 4.4 is from the collection of Shrewsbury Museum and Figures 6.6 and 8.1 are from that of the Yale Centre for British Art, and are reproduced with permission.

CHAPTER ONE

Introduction: trees, woods and history

The trouble with trees

English trees, and especially those growing in rural areas, have become a topic of growing concern not only to foresters and scientists but to all who are interested in the countryside and its wildlife. There is a widely shared belief that we have lost, and are still losing, trees from the landscape at a rapid rate, and that our woods – or at least our ancient, semi-natural woods – have been steadily disappearing over the last half century, either replanted with alien conifers or cleared to make way for farmland, large-scale urbanisation or industry. Wider changes in the global environment have seen woodland devastated by storms, most notably the great gale of 1987. Most worrying of all is the fact that the health of England's trees appears to be deteriorating. The terrible toll taken by Dutch elm disease in the 1970s and 1980s has been followed by a string of further epidemics, including canker and leaf miner in horse chestnut, oak processionary moth, *Phytophthora ramorum* in larch and, most recently and most worryingly, ash Chalara (Cheffings and Lawrence 2014) (Figure 1.1). All are caused by invasive organisms – fungi, bacteria or insects – and have thus been seen as a consequence of globalisation, perhaps compounded by the impact of climate change (Brasier 2008). There are further threats of this kind on the horizon, moreover, including emerald ash borer, pine processionary moth and citrus longhorn beetle. In addition, tree health appears to be suffering a more general deterioration, manifested in the identification of such complex and diffuse conditions as 'oak decline' and 'ash dieback' (Denman and Webber 2009). Both involve a progressive thinning of the crown of affected trees, accompanied by signs of general ill-health, gradually leading to the death of the specimen. An acute variant of the former disease, leading to more rapid death and with debated causes, has also been identified. Our trees are in trouble.

The purpose of this book is not to discuss, in scientific terms, these challenges, but rather to place them within a wider historical context. Only then, we would argue, can we take appropriate action to improve matters – or, in some cases, adopt a more relaxed approach. More specifically, we attempt in the pages that follow to assess how the numbers of trees and the extent of woodland in England have really changed over the last few hundred years, and to explain why. Trees were generally grown for a purpose, and recent changes in their numbers and distribution are not the consequence of mindless vandalism or deliberate hostility to the natural world but rather the result of definable economic,

1.1 A young ash tree suffering from Chalara. This serious disease, caused by a fungus called *Hymenoscyphus fraxineus*, was first noted in England in 2012. It is one of a series of diseases, pests and infections that have appeared over recent years.

social and ideological developments, which we need to explore and understand. An historic approach can also help us to put current issues with tree health in perspective, allowing us to assess, for example, the extent to which trees have always suffered from pests and diseases. This is important because some recent writings appear to assume that, in the past, trees lived in a state of perpetual rude health, untroubled by illness. Historical enquiries can also help us to identify medium- or long-term changes in the environment (in addition to the increased movement of wood, timber and live plant materials around the globe) that might have contributed to declining levels of arboreal health. Above all, such investigations can provide a better understanding of the true character of our tree populations and can help us to assess whether the ways in which these have developed over time – in terms of species composition, management and age structure – might have increased their susceptibility to infection. It is often assumed that the particular kinds of tree we find in the countryside – the ubiquity of ash (*Fraxinus excelsior*) and oak (*Quercus robur* and *Q. petraea*), for example, and the relative rarity of trees such as wild service (*Sorbus torminalis*) – is a consequence of natural factors, but this is only partly true. The key argument of this book is that for centuries the overall numbers of trees, the proportions of different species present in different districts and the ways in which they were managed have all been deeply embedded in social and economic structures. Indeed, some of our current problems may be caused, in part, by assuming that tree populations are more 'natural' than they really are.

It goes without saying that the history of England's trees and woodlands has received considerable attention from other researchers in the past. In the immediate post-war period A.G. Tansley wrote much about the origins of woodland in his monumental *The British Isles and their Vegetation*, while H.L. Edlin, a forester by training, wrote extensively about the uses and management of woods, both traditional and contemporary (Tansley 1949; Edlin 1944; 1949; 1958). More recently, George Peterken, although mainly concerned with the ecology of woodland, has made an important contribution to our understanding of its history, and Charles Watkins has provided perceptive overviews of the development of woods and forests in both Britain and Europe (Peterken 1976; 1981; 1996; Watkins 1990; 2014). Above all, the late Oliver Rackham has written extensively on the subject, effectively dominating the field for many decades (Rackham 1976; 1980; 1986; 2006). Indeed, some readers might be wondering what there is to say on the topic that Rackham has not already discussed. The answer is that, in subtle but significant ways, our focus is different. While his particular interest was in woods, ours is more in trees growing more widely, on farmland and pastures; and while he concentrated on the management of woods and trees in the Middle Ages, our principal concern – as the title of this volume makes clear – is with more recent centuries. Our main interest is in the development of the trees that exist in the countryside today, and most of these are not very old. Some surviving specimens, it is true – mostly old oak pollards – date back to the Middle Ages. But they are relatively rare. For the most part, even our acknowledged 'veteran' trees – ancient examples of their particular species, hosts to a range of rare insects and other organisms and vital for maintaining biodiversity – post-date the start of the sixteenth century. Most trees, of all species, were planted in the eighteenth, nineteenth or twentieth century. This relatively recent character of England's tree population is mirrored by the nature of our sources, for, in a happy congruence, it is only from the seventeenth century that these really provide useful information about trees and their management.

Terms and definitions

It might be useful to begin this enquiry by explaining what we mean by a *tree*, but this is by no means straightforward. There is no clear or accepted distinction between a shrub or a bush, on the one hand, and a 'tree', on the other. Some people count hazel (*Corylus avellana*) as a tree, for example, and it can indeed reach a respectable height in the right conditions, but others will always consider it a shrub. There are, moreover, different definitions of the size or age at which a sapling – often described in the documents as a *samplar*, *standil* or *teller* – becomes a mature tree, and this causes many problems when we attempt, for example, to calculate the density of trees in any area in the past, for our sources often do not tell us the age or size of trees that are being included or excluded from consideration or enumeration, making comparisons with other periods and areas problematic. In broad terms, in the pages that follow we reserve the term 'tree' for those species capable of reaching ten metres or more, and for those individual specimens which have been allowed so to do. The point at which a 'sapling' becomes a tree is more complicated, and depends to an extent on species, but for most trees this occurs at between five and ten years of growth. Such ambiguities and imprecise definitions will accompany all our investigations. They are part and parcel of the world of trees.

Traditionally, trees were managed in two main ways. Some, often referred to as standards or maidens, were left to grow until they had reached a reasonable size and were then felled, processed and used for constructing houses, ships and other things which demanded large pieces of wood or *timber*, as it was usually called. Timber trees were thus different from *pollards*, which were allowed to grow only until they were between two and three metres high, at which point they had their leading shoot removed so that they threw out a number of branches from the same approximate height. They were then repeatedly 'lopped', at intervals of ten or so years, to produce a regular crop of fairly straight, narrow-sectioned pieces of wood, often referred to as *poles*, which were suitable for fencing, fuel and a multitude of other everyday uses. Long after this form of management has ceased pollarded trees still retain a distinctive form, with a relatively short bole and with branches all radiating from a narrow band (Figure 1.2). Once again, the documents on which we depend can mislead and confuse. 'Lopping' can also refer to other forms of management. In post-medieval sources, but more frequently in medieval ones, we hear of the practice of 'shredding', or removing the side branches from a tree, leaving a small tuft on top. This was principally carried out in order to stimulate twiggy, 'epicormic' growth from the trunk of the trees (in oaks especially) which could be harvested and used as winter fodder for

1.2 An ash tree planted in a hedge established when common land was enclosed at Wymondham in Norfolk in 1816. Its growth pattern – with branches all rising from the same level, above a short trunk or bolling – shows that it was originally pollarded.

1.3 Actively managed sweet chestnut coppice at Fernhurst in Sussex. Sweet chestnut is a prominent element in the understorey of many areas of ancient woodland in south-east England, although it is not an indigenous species.

1.4 Actively managed coppice of hazel and bird cherry in Wayland Wood, Norfolk. Today, coppicing is mainly, as here, carried out for nature conservation: the wood is managed by the Norfolk Wildlife Trust.

livestock. Early writers also often talk of the need to lop timber trees – to trim them up to give them straighter stems, or *boles* – and this was done, in particular, with elms. The distinction between trimming a tree for timber and shredding it was thus a vague one, as also in some cases was that between trimming for timber and pollarding. Many a young tree originally intended by the planter for timber might be converted, when scarcely mature, into a pollard, either intentionally or by over-enthusiastic or incompetent lopping.

In spite of such difficulties and confusions, and blurring of categories and definitions, trees as defined above were – at least in broad terms, and in most contexts – quite different from *coppices*, even though many species (including oak, ash, maple (*Acer campestre*), beech (*Fagus sylvatica*), elm (*Ulmus procera* and *U. carpinifolia*) and hornbeam (*Carpinus betulus*)) could be found in both forms. Coppicing is the practice of repeatedly cutting a tree or shrub down to a point at or near ground level on a rotation (Figures 1.3 and 1.4). The *stool*, or base, rapidly regenerates, producing a crop of straight poles. Coppicing is thus, essentially, pollarding at ground level. It is an ancient practice, and much of the material used to construct the Neolithic and Bronze Age wooden trackways preserved in waterlogged peat deposits in the Somerset levels appears to have come from coppices. This form of management was practised in woods, where it was usually combined with the cultivation of timber trees, although these could not be planted too densely or else their canopy shade would suppress the growth of the coppiced plants (the *underwood*) beneath. Coppicing was not, however, restricted to woodland. Many hedges in post-medieval England were, in effect, managed as linear coppices (the word 'coppice' could refer to both the method of management and an area of trees managed by such a method). Wherever it was carried out, coppicing was difficult to combine with grazing, for livestock would browse off the young shoots of the regenerating stools, and harm or even kill them. Where a hedge was coppiced it was carefully protected in the first years of regrowth by excluding sheep and cattle (if possible) from the adjacent land, and by erecting some kind of 'dead hedge' of brushwood along its length, even if this was only a collection of staked thorns. In woods, livestock were generally excluded, at least from areas that had been recently cut, by using similar 'dead hedges', and the perimeter of the wood was usually securely enclosed with a ditch and bank surmounted by a hedge or fence. In northern districts the outer face of the bank might be revetted with stone, or a stone wall used instead of a bank (Jones 2012, 96–7). Trees in woods were not pollarded, except some of those growing on the boundaries. This was because pollarding – in effect, aerial coppicing – was carried out in contexts where the land was grazed by livestock, so that the growing poles had to be raised up out of their reach. Pollards were thus characteristic of pastures and meadows and, in particular, of grazed woodlands, or *wood-pastures*, areas where cattle, sheep or deer grazed on grass or other herbage beneath a variable density of pollards and timber trees (Figure 1.5). As we shall see, pollards were also commonly found growing in hedges, and sometimes in walls, on farmland, although (as perhaps implied by the above account) their presence here has rather different explanations.

1.5 Staverton Park, Suffolk: one of the best surviving examples of wood-pasture, or grazed woodland, in eastern England.

Wood-pastures have largely disappeared from the landscape, so that when we examine the history and natural history of semi-natural woodland today we tend to think of areas managed, or formerly managed, as coppice-with-standards. But grazed woods were extensive in the Middle Ages, especially on common land, and in many districts large tracts survived well into the eighteenth or nineteenth century. Their precise extent is difficult to establish in any period because there is no obvious or accepted point at which a grazed woodland becomes a pasture, thinly scattered with trees. Once again, the study of arboreal history is plagued by problems of definition.

Pollarding is seldom carried out today, except to control the growth of trees in particular circumstances, and while coppicing does occur in many woods this is mainly for conservation purposes, rather than to produce wood. Long-established or ancient woods contain a range of distinctive mosses, fungi, bryophytes, lichens and, in particular, herbs. The latter are the so-called 'ancient woodland indicator species', such as dog's mercury (*Mercurialis perennis*), bluebell (*Hyacinthoides non-scripta*), primrose (*Primula vulgaris*), wood anemone (*Anemone nemorosa*), wood spurge (*Euphorbia amygdaloides*), wood melick (*Melica uniflora*), yellow archangel (*Lamiastrum galeobdolon*) and water avens (*Geum urbanum)* (Rotherham *et al.* 2008, 36–7) (Figure 1.6). All these plants flourish best where coppicing is practised, and where the woodland floor is thus subjected to repeated periods of light followed by increasing amounts of shade during the later stages of the coppice cycle (Colebourn 1989, 70). Their mode of propagation ensures that they spread very gradually from places where they are established, and their presence in a wood (together with evidence of former management by coppicing) is thus generally taken as evidence of its status as *ancient woodland*: a technical, quasi-legal term for woods which have been in existence since at least 1600 (Peterken 1981, 11; Spencer and Kirby 1992). These are afforded particular protection in legislation relating to planning and other matters and are usually considered quite separate from and, in conservation terms, superior to 'recent' woods. While some examples are 'secondary' in character – that is, they have grown up long ago on abandoned farmland or settlement sites – it is agreed that the majority have evolved directly from the natural woodlands which, according to many scientists, covered the country before the arrival of the first farmers. They provide a direct link, that is, with the natural woodlands of remote prehistory.

In fact, several of these widely-held ideas are open to question. 'Ancient woodland indicators' occur not only in old woodland but in long-established hedges, from where they can colonise newly planted woods with remarkable speed. This, together with the fact that new examples of coppiced woodland were established well into the nineteenth century, ensures that the line between 'ancient' and 'recent' woodland is less clear-cut than is sometimes suggested (Stone and Williamson 2013; Barnes and Williamson 2015, 122–32). In addition, most of these 'indicator' plants have poor resistance to grazing and they may thus have been poorly represented in the *grazed* woodlands of

1.6 'Ancient woodland indicators' – bluebell and wood anemone – carpet the floor of a wood in mid-Norfolk.

ancient times, from which coppiced woodlands were enclosed (Rotherham 2012). To a large extent, their prominence in coppiced woods is an artefact of management and a sign that these are essentially *unnatural* environments. But this in turn brings us to the tricky question of the character of the relationship between the trees and woodlands present in the landscape before the advent of farming in the Neolithic and those that have existed in the countryside during the historic period, and especially over the last few centuries.

Transformations of the natural

Many people who enter an area of ancient, semi-natural woodland – whether an outgrown coppice or, perhaps especially, one of the few surviving tracts of grazed woodland, such as Staverton Thicks in Suffolk (Figure 1.5) – probably feel that they are in an environment which has a direct connection with the wild, natural vegetation that existed in England before the advent of farming in remote prehistory – one, perhaps, that resembles the archaic, pre-Neolithic landscape. Indeed, as early as 1839 John Main suggested that 'the greater part of the continent of Europe, as well as its islands, was in an early period almost entirely covered with wood', and that old woods represented the tattered remnants of this original vegetation (Watkins 2014, 180). As we have already seen, this assumption is wrong in the sense that most coppiced woods were intensively managed to produce wood and timber in ways that drastically modified their structure and species composition. But even the wider (and wilder) grazed woodlands, out of which managed woodland was enclosed in the course of the early Middle Ages, had already undergone many changes as a consequence of human exploitation. Indeed, one clear sign of how far we have come from the 'natural' landscape in England is the debate that continues to rage about what precise form this may have taken.

In the middle decades of the twentieth century it was generally assumed that, before the arrival of farming at the start of the fourth millennium BC, the English landscape had been characterised by dense and almost continuous closed-canopy woodland, dominated by oak. This had developed following the end of the last Ice Age, around 11,000 BC. The land was gradually colonised by plants as the temperature warmed and as a continued connection with continental Europe allowed them to move northwards with relative ease, leading in time to the development of this climax vegetation of dense forest. There were good reasons for believing this model. The emerging science of palynology – the study of pollen grains preserved in waterlogged conditions, free from the presence of oxygen, and thus protected from decay and decomposition – suggested that woodland had been extensive before the advent of farming. It also showed the dramatic signature left by the latter development, in the form of a sharp decline in the amount of pollen produced by tree species and a concomitant increase in that of plants associated with arable fields. Moreover, it had long been observed that if land is left derelict for any length of time the predicted succession to woodland begins once again: within a short period grasses and herbs will give way to scrub, and scrub to trees. As early as 1880 experiments carried out at Rothamsted in Hertfordshire showed how the deliberate abandonment of two small plots of land led to the rapid regeneration of woodland dominated by oak and ash (Lawes 1895; Brenchley and Adam 1915).

The idea that oak (*Quercus* sp.) was the most important tree throughout the country in these primeval woods also made a lot of sense given that it was, and is, the most common timber tree found in ancient woodland in England, and one of the most common in the wider countryside. In the 1950s Edlin thus believed that: 'Oakwoods of one kind or another are so ubiquitous over Britain, that one can advance, fairly safely, the working hypothesis that mixed oakwood is, or has once been, the normal forest cover on most areas that can carry woodland at all' (Edlin 1958, 74). However, as techniques of pollen analysis improved it became apparent that across most of lowland England oak had not, in fact, been the most common tree in the 'wildwood' at all, but rather small-leafed lime (*Tilia cordata*), accompanied by varying mixtures of oak, hazel, ash and elm, and with pine (*Pinus sylvestris*) and birch (*Betula pendula*) locally important (Rackham 2006, 82–90). Only in the north and west of the country had oak been dominant, although even here significant amounts of hazel, birch, pine, alder (*Alnus glutinosa*) and elm had also been present (Bennett 1988, 251; Rackham 2006, 82–90). Lime is now a relatively rare tree in most parts of England, both in woods and elsewhere, with only localised concentrations, especially in parts of Essex, Suffolk, Lincolnshire and the West Midlands (Pigot 2012, 59). This makes its prehistoric dominance in, for example, the area around London even more striking (Greig 1989). Quite why it was replaced by other species remains uncertain, but there is no doubt that the traditional dominance of oak in old woods is the outcome of economic factors as much as, if not more than, environmental ones. Oak makes excellent timber. Lime, in contrast, has only limited uses.

Established ideas about the character of the natural, pre-Neolithic vegetation were, however, more dramatically challenged at the end of the twentieth century by the Dutch ecologist Frans Vera. Already, in 1986, Oliver Rackham had suggested that some areas of more open ground must have existed as part of the country's natural landscape, to provide a home for the vast number of non-woodland, open-country species which today dominate our flora. But Vera took such ideas much further. He argued that close inspection of the pollen evidence showed that a number of the species well-represented in the pre-Neolithic landscape are today usually out-competed in dense woodland; some, such as blackthorn (*Prunus spinosa*), hawthorn (*Crataegus monogyna* and *C. laevigata*), rowan (*Sorbus aucuparia*) and apple (*Malus sylvestris*), are usually considered characteristic plants of the woodland edge (Vera 2002, 92). Wood-edge herbs such as nettle (*Urtica dioica*) and sorrel (*Rumex acetosa*) also figure prominently in early pollen records. All this suggested a more open landscape than had previously been assumed. Vera went on to theorise that full succession to dense and continuous woodland had been arrested by the presence of large herds of wild herbivores, especially auroch (wild cattle), horse and deer: the landscape thus resembled savannah more than 'forest'. In part the apparent dominance of trees in the landscape, suggested by the pollen evidence, was an illusion resulting from the fact that intensive grazing, as well as suppressing tree growth, would also have limited the extent to which grasses and other open-country herbs could flower, thus hiding their ubiquity; while many trees – oak and lime among them – actually emit more pollen when growing in open situations than they do in woods (Vera 2002, 88). There were other important strands to Vera's arguments. The primeval landscape, he suggested, had not been entirely open, with only a light scattering of trees. There were also patches of scrub, composed of hawthorn and blackthorn, where larger collections of trees could seed and grow out of reach of grazing animals. Small patches of woodland thus developed in places, but they were essentially transient. When the trees reached maturity they shaded out the protecting thorns beneath, and as they died (perhaps because they were damaged by livestock, perhaps through old age) the area they occupied reverted to open grassland. Driven by the grazing of large herbivores, the landscape was dynamic and in a constant state of flux.

Vera's ideas have been particularly influential among those naturalists and ecologists who believe that the future maintenance of biodiversity in Europe is best assured through a policy of 'rewilding': that is, by the reduction or removal of human influence from extensive tracts of land in order to allow these dynamic grazed landscapes to return (Soulé and Noss 1998; Foreman 2004; Monbiot 2015). The creation of large 'rewilded' areas was first advocated in the United States, but the concept has gained ground steadily in Europe and the UK over recent decades. It has been put into practice at Oostvaardersplassen in the Netherlands and in England at Ennerdale in Cumbria and on the Knepp Castle estate in Sussex (Foreman 2004; Jepson 2015; Soulé and Noss 1998; Monbiot 2015). Ideas about 'rewilding' did not arise directly from Vera's theories, but have received much support from them, and proponents use Vera's 'dynamic savannahs' as their model for what these re-wilded tracts should be like. It must be emphasised, however, that many ecologists have doubts about Vera's ideas, some expressing uncertainty over whether herbivore numbers would indeed have been enough to prevent substantial woodland regeneration (Hodder *et al.* 2009; Kirby and Baker 2013). The loss of the west European 'megafauna', including the elephant, through the great extinctions of the late Pleistocene would, in particular, have much reduced the impact of mammals on the natural flora (Yalden 2013). These extinctions were themselves the consequence, at least in part, of human predation (Lyons *et al.* 2004), and as the 'wildwood' was developing in the immediate post-glacial period Mesolithic hunters were better armed than before and present in greater numbers. It seems likely that human predation would have kept the numbers of auroch, horse and other large herbivores severely in check. While it is almost certainly true that the pre-Neolithic landscape of northern Europe was more varied than we once assumed, and probably included some quite extensive areas of open ground, the majority of the land surface was probably occupied by closed-canopy forest. At the very most, as Samojlik and Kuijper put it, 'The prehistoric landscape in Europe most likely consisted of large stretches of closed high forest dominated by browsing ungulates, interspersed by open or part-open landscapes dominated by large herbivores' (Samojlik and Kuijper 2013, 157).

Even if there was much truth in Vera's arguments, this does not mean that there is a direct link between the grazed landscapes of prehistory and the woods – grazed or otherwise – that existed in the landscapes of medieval or early modern England. The density of domestic animals in the landscape following the adoption of farming was much higher than that of wild grazers before it (Yalden 2013), and the wood-pastures of the historic period were more intensively exploited than any landscape that had existed in the Mesolithic. Centuries of exploitation must have ensured many significant changes in terms of structural character and species composition – including the marked decline in the importance of species such as small-leafed lime, which we have already noted.

Farming began to replace an economy based on hunting and gathering in England in the early fourth millennium BC. The new way of life was initially brought by immigrants from Europe, but was then perhaps emulated by the indigenous population. The principal crops – early varieties of wheat and barley – were introductions from abroad, as were sheep, goats and probably pigs and cattle (the wild ancestors of both pigs and cattle existed in England, as elsewhere in Europe, but the available evidence suggests that they were not independently domesticated here) (Yalden 1999, 95). The adoption of farming soon led to the development of tracts of ground that were permanently cleared of trees. The chalk downland around Winchester, for example, appears to have been largely deforested by the middle of the fourth millennium BC (Watson 1982, 75–91), while in the Peak District blanket bog, which had already begun to form following limited woodland clearances during the Mesolithic, had spread to something like its present extent by 3000 BC (Tallis 1991, 401–15). The rate of deforestation probably intensified in the middle Bronze Age, during the second half of the third millennium BC, and by the end of the Iron Age archaeological surveys suggest that settlement was widespread on almost all soils, including some of the heaviest clays (Fowler 1983; Pryor 1998). In the Roman period, according to many researchers, the landscape was already 'something like the present countryside, farmland with small woodlands rather than woodland with small clearings' (Yalden 1999, 128). Nevertheless, in general, even Roman settlements were smaller and more scattered than those of the Middle Ages, and there are good grounds for believing that many tracts of woodland continued to exist, and that in certain areas these were extensive and continuous, as, for example, in what were to become Blackdown Forest or the Forest of Dean (Dark 2000, 81–129; Straker et al. 2007, 145–50; Brown and Foard 1998, 67–94). The area of woodland probably increased significantly in many districts as the population declined once again at the end of the Roman period. In spite of a gentle recovery of population from the seventh century, and an associated re-expansion of the cultivated area, woodland still remained extensive in many regions at the time of the Domesday Survey of 1086, by which time there were perhaps two million people living in England.

As documentary sources become available in the course of the Anglo-Saxon period they confirm that most woodland was intensively exploited, not only for grazing but also as a source of wood and timber. Particular areas were already allocated to the use of particular communities and individuals, and were protected from illegal exploitation. As early as 690 a law of King Ine of Wessex laid down that:

> If anyone destroys a tree in a wood by a fire and it becomes known who did it, he shall pay a full fine; he shall pay 60 shillings, because fire is a thief. If anyone fells a large number of trees in a wood, and it afterwards becomes known, he shall pay 30 shillings for each of three trees. He need not pay more, however many there may be, because the axe is an informer and not a thief. If, however, anyone cuts down a tree that can shelter thirty swine, and it becomes known, he shall pay 60 shillings (Whitelock 1955, 365).

Many early medieval sources emphasise the particular importance of woods as places of pannage – that is, where pigs were pastured in the autumn, fattening on the harvest of nuts, acorns and mast. The will of the noblewoman Æthelgifu, drawn up around 970 and dealing with far-flung estates in Hertfordshire, Bedfordshire and Northamptonshire, bequeathed a herd of swine 'and the swineherd with it' at *Achurst* in Aldenham in Hertfordshire, another herd at *Langford*, a herd and its swineherd at Standon and what appear to have been two separate herds at Gaddesden, while land at Tewin was given 'as swine-pasture' (Crick 2007). But other kinds of livestock were also grazed in woodland. We tend to think of sheep as animals that live primarily off grass, but this is largely because that is what we feed them on; in fact, they will happily consume woodland vegetation, being particularly partial to ash, ivy and holly. Cattle will likewise browse off whatever foliage they can reach. As we shall see, both continued to be fed on branches cut from ash and holly trees, in particular, well into the post-medieval period. The felling of trees and the cutting of shrubs, the grazing of herds and flocks and the disturbance created by pigs ensured that the woodlands of Anglo-Saxon England were already a long way from the archaic 'wildwood', whatever precise form this had taken. By the twelfth and thirteenth centuries the pressure on woodlands must have been considerable, not only in this country but across wide areas of Europe. It has been suggested, for example, that the species composition of the woods in the Carpathian basin changed significantly during the Middle Ages, towards an increased dominance of oak, because of the importance of acorns as winter feed for pigs: as early as this, it is argued, deliberate selection or even replanting was being undertaken (Szabo 2013).

In England, the extent of the wooded 'wastes' continued to contract in the period after Domesday, as the population reached unprecedented levels, as more land was converted to farmland and, perhaps, as portions degenerated to open pasture under the pressure of grazing. From the eleventh century, and on an increasing scale from the twelfth, some of what remained was enclosed and converted into coppice woods by manorial lords, so that wood and timber, protected from grazing beasts, could be produced in greater quantities. But grazed woods, or wood-pastures, continued to exist right through the Middle Ages and into the period covered by this book. Most were exploited by local communities as areas of common land, although the lord of the manor was the legal owner. Others, as we shall see, were (like coppiced woods) private land, enclosed with stout fences raised on substantial banks. These were the deer parks – special enclosures in which deer were kept, for hunting and to supply venison for the owner's table – most of which, unlike the landscape parks that came later, lay at a distance from the owner's residence because they had been created out of surviving tracts of wooded 'waste' on the periphery of agricultural territories.

Regions

Trees and woodlands cannot be studied in isolation from the wider landscapes of which they are a part, and in medieval and early post-medieval England the pattern of fields and settlements displayed an almost infinite degree of variation. As several generations of landscape historians and others have emphasised, however, a useful starting place – as long as we remember that it is *only* a useful starting place – are the two distinctions between the 'highland zone' in the north and west of the country, where a high proportion of the land area lies at a height of more than 250 metres and is occupied by moorland, and the lowlands; and, within the lowlands, between what can be described as 'woodland' and 'champion' regions (Williamson 2013a, 125–46; Roberts and Wrathmell 2002). 'Champion' regions were characterised by large villages which, albeit to a decreasing degree over time, farmed extensive communal 'open fields'. The village farms consisted of numerous unhedged strips, each around seven metres wide, that were scattered with varying degrees of regularity through the territory of the township. For purposes of cropping, the strips ('lands' or 'selions') were grouped into bundles called 'furlongs' or 'shotts', and these in turn into two or three great 'fields'. One of these lay fallow each year and was grazed by the village livestock, the dung from which replenished some of the nutrients depleted by repeated cropping (Hall 1982; 1995). Farming was organised on highly collective lines, so that (for example) each year the same crop was cultivated on all the strips in a particular furlong, and communal regulations governed the exploitation of the 'wastes' or common land of the village. In many places these communal farming systems survived into the eighteenth or nineteenth centuries, when they were removed by large-scale enclosure, often through parliamentary acts. Thus was created the landscape of straight-sided fields that is often referred to – following Rackham – as *planned* countryside (Rackham 1976, 2–3; 1986, 4–5). 'Champion' landscapes could, to a degree, be found throughout the lowlands, but they were concentrated in a broad band running from Yorkshire through the Midlands and to the south coast (Figures 1.7 and 1.8) (Rackham 1986, 4–5; Williamson 2013a, 125–46). They were thus to be found in the north of England as well as in the south. The Vales of York and Mowbray, for example, were classic landscapes of nucleated villages

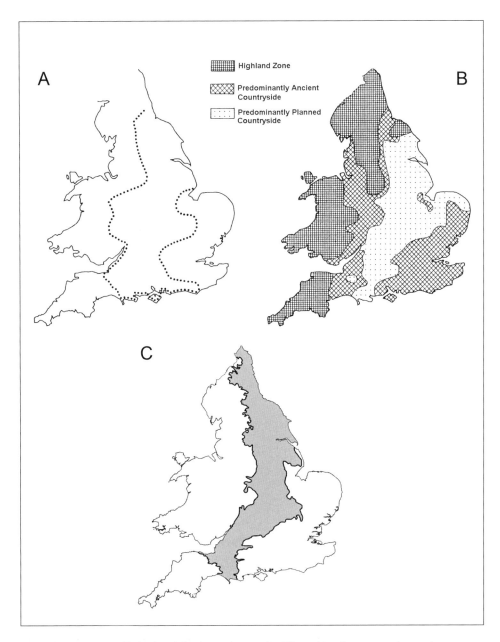

1.7 Landscape regions in England, as defined over the years by different historians, geographers and archaeologists. (a) shows how Howard Gray in 1915 defined the boundaries of his 'Midland System', comprising highly communal open-field systems farmed from nucleated villages; (b) represents Oliver Rackham's distinction between the 'planned' countryside, created by the late enclosure of open fields and commons, and the 'ancient' countryside or 'woodland' landscapes, of dispersed settlement and old enclosure; (c) is the 'Central Province', as defined by Roberts and Wrathmell (2002) – essentially, the 'champion' landscape of extensive open fields and large villages. Although there are significant differences between these various maps – which are not, in fact, all mapping quite the same thing – there are also clear similarities. The 'Champion' belt of villages and large open fields, generally enclosed late in the post-medieval period, runs in a broad band down the centre of England, separating zones characterised by early enclosure and more dispersed settlement.

and extensive open fields, as were the chalk wolds of Yorkshire and Lincolnshire. Even in parts of Durham and Northumberland, landscapes of this type could be found.

To the south, east and west of this tract of 'champion' landscape lay what early topographers described as 'woodland' districts, and which many modern writers, following Rackham, often term *ancient countryside*. In these, settlement was more scattered or dispersed in character, with numerous isolated farms and small hamlets (many of them clustered around areas of common land) as well as – and in some districts instead of – nucleated villages. In some of these areas the cultivated land was always held, from the earliest times, 'in severalty' – that is, in individual occupancy – and the landscape comprised networks of hedged or walled fields. But in many such districts open fields existed. In some cases they occupied only a part of the cropped area, which was otherwise occupied by enclosed fields, but they sometimes embraced most of the cultivated land (Campbell 1981; Roden 1973). Either way, they were usually laid out and organised in rather different ways to the great fields of the 'champion'. The holdings of individual farmers, rather than being scattered throughout the lands of the township, tended to be clustered in particular areas of the fields, usually near their farmstead, and communal controls on the organisation of agriculture were usually less pervasive. 'Woodland' countryside, like 'champion', was found in the north of England as much as in the south; it was particularly extensive, for example, on the western flank of the Pennines, in Cheshire and Lancashire.

The origins of this broad division in the landscape of lowland England, a matter of continued debate among landscape historians, need not detain us here. What is important is that 'woodland' and 'champion' are simplified terms which in reality cover a range of landscapes. There were, in particular, two main types of 'champion' countryside (Williamson 2013a, 125–46). On light, well-drained land, often overlying chalk or sand, such landscapes usually included extensive tracts of unploughed ground – downland or heath – in addition to the open arable. These were the classic areas of 'sheep–corn' husbandry, in which large flocks were grazed on extensive pastures by day and by night kept on the arable in folds made of wooden hurdles, so that they dunged the land (Kerridge 1993, 25–30). In Midland districts, by contrast, champion landscapes were mainly associated with heavy clays, and generally lacked the great open pastures – the nutrient reservoirs of down or heath – of the sheep–corn lands. Heavy soils retain nutrients better than light, permeable ones, and close-folding for much of the year would have damaged the soil structure. The fallows were dunged by the village livestock, but in a less intensive manner (Kerridge 1993, 77–9). It is also important to emphasise that the distinction between the 'two countrysides' can be too neatly drawn. In reality, many districts displayed intermediate characteristics, and there was often much blurring of the boundaries.

This said, the woodland/champion distinction is important in the context of this study because it had a determining influence on the density of trees. 'Woodland' districts

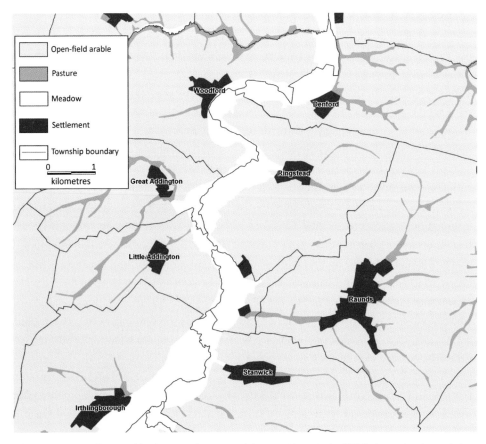

1.8 The landscape of eastern Northamptonshire, around the river Nene, in the Middle Ages, showing typical 'champion' countryside of near-continuous open fields, large meadows and nucleated villages.

were, on the whole, better endowed with woods than champion ones, but their name seems to have been coined by early topographers because of the large amounts of wood which were provided by the hedges and hedgerow trees. Even where open fields were prominent they were often small, numerous and individually hedged. In the 'champion', in contrast, hedges and therefore hedgerow trees were fewer although, as we shall see, these landscapes were not quite as devoid of trees as has sometimes been suggested. And we should also note that even before the great wave of parliamentary enclosures of the later eighteenth and early nineteenth centuries large areas of open fields were removed by piecemeal consolidation or more formal legal processes. The smaller open fields of the 'woodland' were particularly affected, but even in 'champion' districts hedged closes proliferated, especially in areas of poorly draining clay where much land was, in the period from the fifteenth century, laid to grass. By the time of the first parliamentary enclosures even counties such as Northamptonshire or Leicestershire, classic 'champion' lands, had lost around half their open arable (Williamson *et al.* 2013, 133–43).

Trees and property

As we have already intimated, the management of trees and woods in England was, by the late sixteenth century, firmly embedded in the wider structures of social and economic life. Modes of management, even the particular species found in different locations, were a consequence of human choice. So far as the evidence goes, this had been the case for centuries. Trees were somebody's property, but not always in a simple or absolute sense. Tenants, whether holding by medieval customary tenures and copyholds or by more modern leases, did not usually have the right to fell the timber trees on their farms: these were reserved to the manorial lord, or the landowner. Farmers thus regarded hedgerow trees with some hostility, for they tended to interfere with the practice of farming, their roots robbing the soil of moisture and nutrients, their shade reducing crop growth. Most tenants wanted few trees other than pollards, from which they did normally derive some benefit. In most cases, while the trunk, or *bolling*, of the pollard belonged to the manorial lord or freehold owner, tenants had the right – by custom, or by the terms of their leases – to crop the poles. The same general rule usually, although by no means always, applied to the trees growing on common land. In addition to this, different forms of woodland management were associated with different patterns of ownership. Coppiced woods were thus invariably private property: in the medieval period they formed part of the demesnes of the manors in which they lay, while in the post-medieval period they were freehold property, owned by members of the gentry or aristocracy as part of some larger estate. In some cases, common rights might be exercised within such woods by local people – the right to gather fallen wood or, perhaps, to graze limited numbers of stock when the coppices had regrown sufficiently. But these rights were usually limited and carefully policed. Common woods (or wooded commons) certainly existed, but these were invariably wood-pastures – grazed woods – without a coppiced understorey. The private status of coppiced woods, and the fact that they were cut out of wider tracts of common 'waste' in the early Middle Ages, may help explain the character of their boundaries. Medieval woods were usually enclosed by particularly substantial boundaries, usually in the form of banks topped by hedges or fences and rising as much as two metres above an external ditch – much larger than the hedges and banks employed to contain livestock in other contexts. This suggests that they also had a symbolic function, as symbols of lordly possession, and that they were intended to deter the sheep and cattle deliberately introduced by trespassers keen to exercise ancestral rights as much as to prevent the entry of animals acting on their own volition.

One term relating to woods and their management that continues to cause some confusion, and which is again closely tied to patterns of ownership and rights, is 'forest'. In the Middle Ages forests were not necessarily densely wooded areas – some, in the highland regions of England, contained very few trees. They were simply tracts of land subject to forest law, a bundle of rules and regulations intended to preserve deer for the king's hunt: a 'chase' was similar, but here the deer were usually preserved for the benefit of an aristocratic family. People living in forests were obliged to keep their hedges and fences low, so as not to interfere with the free movement of deer; they could not forcibly remove deer from their growing crops; and they could keep dogs only if they were deliberately lamed. There were also limits placed on the amounts of new land that could be 'assarted', or brought into cultivation, within forests. In lowland districts most forests possessed a wooded 'core', but beyond this there was a much wider penumbra of land over which forest law applied. This comprised farmland, even entire villages, and was defined in different ways by changing 'perambulations' of the forest bounds over the centuries: indeed, at one stage in the twelfth century, around a fifth of the country's land area was technically 'forest' (Page 2012). The wooded heart of the forest comprised two main elements: coppices and plains. The coppices were enclosed, and either private or royal property, and they were managed as most coppices were, except that the fences were usually removed and deer allowed in at a late stage in the rotation. The plains, which were usually owned by the king but had the status of common land, were wood-pastures in which deer grazed alongside livestock owned by commoners from surrounding villages. Some common rights were also usually exercised over the enclosed coppices, especially the right to collect fallen wood. Forests often included other characteristic features. One or more 'lodges', which served as accommodation for the officials responsible for administering the forest, usually existed and there were often some royal parks, enclosed from the wider forest, where deer could be more carefully preserved. Forests were found in many parts of the country, but they were a particularly important aspect of the landscape in 'champion' regions, where forests such as Bernwood or Rockingham represented substantial islands of well-wooded ground in landscapes that were otherwise dominated by arable open fields and had relatively little woodland.

Patterns of rights and landownership might, in some areas at least, structure the distribution of different types of woodland, probably more so than aspects of soils and drainage. Both coppiced woods and commons (wooded or otherwise) were most extensive on the more agriculturally 'difficult' soils – acid gravels or poorly draining clays. But where such soils cover large areas coppiced woods were and are noticeably concentrated towards the margins, close to the junction with better agricultural land. In the Weald of Kent and Sussex, for example, Kenneth Witney has observed how, by late medieval times, most coppiced woodland was to be found around the district's periphery rather than towards its centre (Witney 1998). The twelfth and thirteenth centuries saw a steady decline in the economic importance of the Wealden woods as swine pastures and grazing grounds and a concomitant increase in the demand for, and thus in the value of, wood and timber. Areas of woodland were thus enclosed, and more intensively managed as coppices, in places where their products could be transported to markets with relative ease. In the 'central core of the Weald … heavy loads were almost undisposable' because of the difficulties involved in moving laden carts in wintertime along clay roads (Witney 1998, 20). During the great phase of medieval population expansion in the eleventh, twelfth and thirteenth centuries colonisation was thus directed into the more remote districts, where the wooded 'wastes' were progressively felled and turned to farmland or degenerated to open commons.

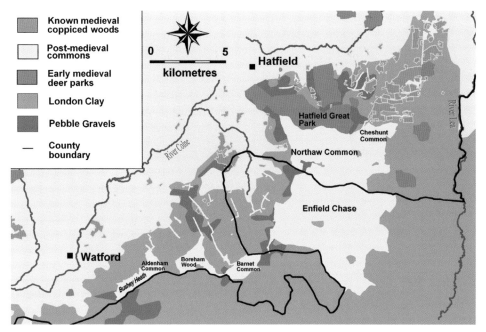

1.9 The 'doughnut' of woodland in south Hertfordshire, as it existed in the eighteenth century. Areas of enclosed, coppiced woodland tend to be found on the margins of the poor London Clay and Pebble Gravel soils close to the major valleys. The higher, poorer ground is occupied by common land, most of it comprising wood-pasture in the Middle Ages.

Woodland, therefore, survived mainly on the periphery of the Weald. In East Anglia and Essex coppiced woods similarly cluster towards the edges of the valleys dissecting the extensive boulder-clay plateau (Warner 1987, 5–9), and on the poor London clay uplands of south Hertfordshire the same pattern is repeated: woods tend to be found towards the margins of the valleys of the river Lea and Colne; commons (most of them tree-covered until well into the post-medieval period) were found on the higher and more remote ground (Figure 1.9) (Rowe and Williamson 2013, 127–30).

This tendency of enclosed woods to occupy the peripheries of marginal land is often obscured by the complexity of topography. But the 'doughnut of woodland', to coin a phrase, is a widespread phenomenon. A rather similar spatial pattern can be found in many upland districts, although here the causes were slightly different. In many northern and western areas the higher ground, beyond the limits of cultivation, had once been tree-covered, but had degenerated to open moorland at various times before the Middle Ages. In others, and especially further north, trees had often failed to take root to any significant extent owing to altitude, thin soils and exposure. Either way, the treeless character of the upland moors was well established by the start of the period studied here. On lower ground – in the principal vales and dales – woodland was gradually cleared to make way for farmland, a long process with various reverses which continued through the prehistoric, Roman and medieval periods. By the late Middle Ages both woods and wood-pastures tended to be

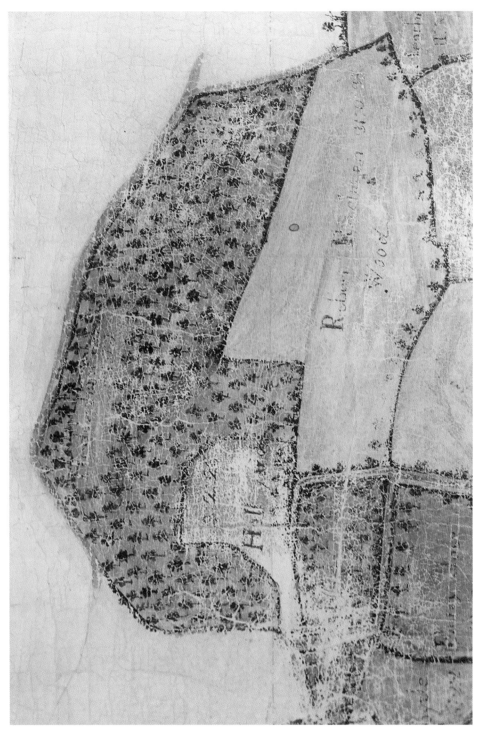

1.10 Hedenham Wood in Norfolk, as depicted on Thomas Waterman's map of 1617. A small cleared area in the centre of the wood, approached from the east by a road, is marked as 'Hall Yards'.

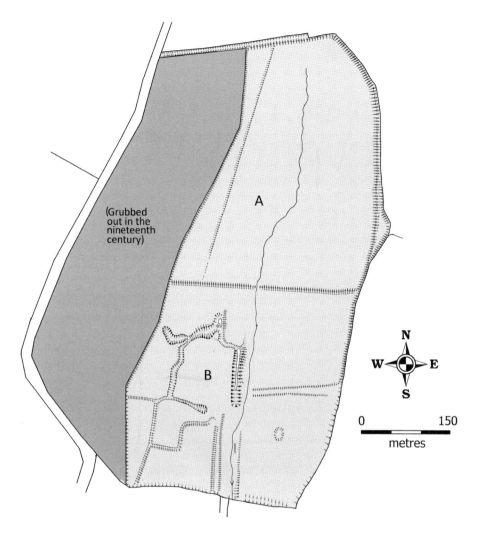

1.11. Hedenham Wood in Norfolk today. The western half of the wood was grubbed out in the mid-nineteenth century, but the archaeological evidence indicates earlier changes. The northern section of the wood (A), devoid of earthworks, is 'primary' in character: the original boundary can be seen running east–west across the middle of the wood. In the fifteenth century the wood expanded southwards, across a network of small fields and the site of Hedenham Hall (B).

concentrated in the area between these two zones, on the sloping ground encircling the main upland masses.

Prelude: woodland and landscape in the fifteenth and sixteenth centuries

Our main focus in the pages that follow is on the period after *c*.1600, but a brief sketch of developments during the previous two centuries or so is necessary in order to set the scene. Across much of England the enclosure and definition of coppiced woods, and their divergence from grazed woodland, occurred within the context of a steady increase in population that continued through the twelfth and thirteenth centuries. Demographic expansion was then brought to an abrupt halt by poor harvests and climatic deterioration, and put into reverse by the arrival of the Black Death in 1348/9, which, with repeated epidemics later in the century, reduced the population by around 30 or 40 per cent. Archaeological evidence suggests that there was then a measure of woodland regeneration: a significant number of small ancient coppiced woods overlie the earthworks of medieval settlements or field systems, while many larger woods partly do so (Barnes and Williamson 2015) (Figures 1.10 and 1.11). For the most part, however, land continued to be farmed as arable or pasture, but with a marked expansion in the latter at the expense of the former. In a less crowded world, rising disposable incomes provided a stimulus to new forms of agriculture and to other aspects of the economy. Population growth resumed through the fifteenth and sixteenth centuries, but a number of the economic and social changes begun by the Black Death continued with little interruption, many of which had important implications for the management of trees and woods.

One of these was the emergence of larger and more specialised farms. Some degree of agrarian specialisation had existed in the twelfth and thirteenth centuries, but most medieval peasants concentrated on the production of the grain upon which, in the last analysis, their survival depended, and even the large 'demesne' farms operated by major landowners were usually grain-factories, supplying urban markets. Agrarian specialisation increased after the Black Death, however, as average farm size began to rise and as the economy generally became more complex and sophisticated (Thirsk 1987; Kerridge 1967). To some extent this was a consequence of tenurial changes. Before the Black Death customary tenants – villeins and cottars – had owed rents and services to a local lord, but farms passed by inheritance within peasant families and in many cases customary land could be bought and sold, the rents and obligations attached to it simply passing to the new proprietor. Only the home farm of the manorial lord – the 'demesne' – was his absolute property, as we would understand it. Through the fourteenth and fifteenth centuries peasant tenures evolved into a range of 'copyholds', some of which effectively recognised local lords as owners of their manors and their tenants as tenants in the modern sense; others provided farmers with a greater degree of security, and some gave them so many proprietorial rights that they effectively joined the ranks of the small numbers of freeholders who had always existed among the peasant population. Such yeoman farmers, as well as many of the gentry and aristocracy, sought ways to increase their profits and the size of their properties as a more complex and market-orientated economy developed.

Many districts continued to focus on grain production, but others came to specialise in livestock farming – in dairying or the rearing or fattening of sheep and cattle. Distinctive agricultural regions thus developed, so that by the start of the period studied here substantial areas of both south-eastern and western England were devoted to pastoral farming, with three-quarters or more of their land under grass. They were mainly 'woodland' districts, where much farmland had always taken the form of enclosed fields or comprised open fields of 'irregular' form that could be enclosed with relative ease in a gradual, piecemeal fashion (Yelling 1977, 11–29; Williamson 2000, 68–9). In the champion Midlands, in contrast, although the soils were often well suited to livestock farming, the large and complex open fields were more difficult to enclose. Extensive areas were laid to grass, but arable continued to dominate the landscape, at least in the period up to 1650. A more commercial approach pervaded many other aspects of rural life and helped to structure distinctive farming landscapes in a host of other ways. Some farmers thus adopted fruit-growing and also the production of cider – especially those dwelling in the west Midlands and the Welsh border counties but also, as we shall see, in parts of south-eastern England – as important side-lines. And in the latter district the cultivation of hops for the London brewing industry became a major concern following their introduction at the start of the sixteenth century. Both pursuits had important implications for the character of farmland trees and woodland management.

During the fifteenth and sixteenth centuries the economy grew more complex and diverse in other, non-agrarian ways. In particular, there was steady growth in a number of key industries, and especially those associated with the mining and processing of iron and non-ferrous metals such as tin, copper and lead. It was often suggested in the past that the expansion of the iron industry, in particular, led to the large-scale destruction of woodland, especially in southern England, for vast quantities of wood and charcoal were required for smelting and forging. But, as Rackham observed long ago, far from 'using up' woodland, the demands of industry ensured its profitability and thus its continued existence (Rackham 1976, 84–6). Wood was not only used as a fuel, moreover. It was in high demand for building industrial premises and infrastructure, and was a raw material in the production of commodities such as gunpowder, while bark (from oak trees especially) was employed in large quantities in the leather tanning industry.

The impact of industrial expansion on woodland in the fifteenth and sixteenth centuries was greatest in parts of northern England. Here, because population levels had remained lower than in the south, wood-pastures had continued to be extensive and coppices of relatively limited extent. Gledhill has analysed over 900 separate twelfth- and thirteenth-century references relating to 400 separate woodlands in North Yorkshire, of which 121 can clearly be identified as wood-pasture or coppice: of these, around 70 per cent were clearly wood-pasture (Gledhill 1994, 168). Only in the period between 1320 and 1400 do references to coppicing start to become more prevalent. Given that this was a period of declining population, the adoption of more intensive management probably indicates growing industrial markets, although rising disposable incomes and thus an ability to consume larger quantities of domestic fuel might also have been a factor. In Nidderdale in the Yorkshire Dales Ian Dormor has detected a similar expansion of coppicing during the later fourteenth and fifteenth centuries. This was in part a response to the degeneration of the local wood-pastures caused by the intensive grazing of the stock kept by Fountains and Byland Abbeys, the monks keen to capitalise on the rising market for meat and wool; but in part, once again, it was a consequence of the growth in local industry. The produce of coppices was used to make charcoal for smelting, or was employed directly for this purpose. By 1450 the area under wood-pasture in Nidderdale had decreased by half (Dormor 2002, 18), and by the time of the Dissolution of the Monasteries in 1539 69 per cent (571 acres, 231 ha) of the abbey woodlands were managed as coppice (Dormor 2002, 54).

The growth of industries in the fifteenth and sixteenth centuries thus raised the value of trees and woodland and intensified their management. But the same period also saw a remarkable expansion in the scale of coal mining and the beginning of the replacement of wood by coal as fuel in a range of manufacturing processes, such as soap production, and for the forging of iron and other metals – although not, as yet, for smelting them. Coal was also employed on an increasing scale as a domestic fuel. Even though the costs and difficulties of transport ensured that its use was geographically restricted – to the coalfields themselves, or to places located on coasts or navigable rivers – coal was already, according to most historians, the largest provider of thermal energy in England by the early or middle decades of the seventeenth century (Wrigley 2010; Warde 2006, 32–9, 67). As we shall see, its use increased steadily thereafter. Of particular significance was the recognition – by Abraham Darby at Coalbrookdale in Shropshire at the start of the eighteenth century – that coke could be used instead of charcoal in the smelting of iron. Even more important was the gradual improvement in transport infrastructure through the eighteenth and nineteenth centuries, which allowed coal to be transported more easily to distant markets, where it might be burned on domestic fires. Eventually, this lowered the value of wood as fuel and led to radical changes in the management of trees and woodlands. But the relationship between coal mining and woodland exploitation was a complex one, for mines needed vast quantities of wood and timber to operate and thus often served as major markets for woodland produce. The same was true of other extractive industries. As Bailey noted of the woodlands in County Durham in 1810: 'the produce in the eastern parts is chiefly applied to various uses about the collieries, as pit props, wagons, waggonways &c.&c. In the western parts the application is mostly for the lead mines' (Bailey 1810, 188).

As we shall see, the expansion of industry was only one of the developments that, beginning in the wake of the Black Death, intensified through the seventeenth, eighteenth and nineteenth centuries. Farm size continued to increase and, in particular, more and more land was acquired as freehold property by the gentry and aristocracy, and held in larger and more continuous blocks. The pace of enclosure also quickened and, from the end of the eighteenth century, increasingly affected areas of common land – including forests – as well as open fields. All these things, as we shall see, were to have a determining influence on the character and management of England's trees and woodlands.

Studying trees

A wide range of sources exists that allows us to chart the history of trees and woodlands in England in the period after *c*.1600. As the estates of the landed gentry became more consolidated and more professionally managed, they generated a mass of documentation: leases, estate surveys, timber accounts and correspondence all provide much detailed information about the numbers, character and management of trees both in woods and on farmland, as do a rather smaller number of early maps. From the late nineteenth century, moreover, Ordnance Survey maps, government records and – from 1946 especially – vertical aerial photographs provide data at a national level, albeit relating mainly to tree numbers and woodland area, rather than to their management. In addition, beginning in the sixteenth century, but on an increasing scale from the late seventeenth century, a series of published texts appeared that were devoted to aspects of land management. Studies of woodland history tend to focus on those volumes specifically devoted to the planting and management of trees, such as John Evelyn's *Sylva* of 1664 or Moses Cook's *The Manner of Raising, Ordering and Improving Forest Trees* of 1676. But equally useful information, especially about the management of non-woodland trees, is provided by texts whose main focus was on the practice of agriculture, such as Thomas Hale's *Compleat Body of Husbandry* (1756), John Worlidge's *A Compleat System of Husbandry and Gardening* (1660) and John Mortimer's *Whole Art of Husbandry* (1707). For the late eighteenth and early nineteenth centuries the *General Views* of the agriculture of the various English counties (some appearing in two successive editions) also tell us much about the management of trees, hedges and woods. By this stage, moreover, writings on forestry begin to take on a more scientific guise and provide more information about tree health – and ill-health – although only from the end of the nineteenth century do really informative texts on these matters begin to appear.

It must be emphasised, however, that on a number of topics our data is limited, biased or misleading. For example, in some cases those responsible for undertaking surveys of the trees present on a property made clear the purpose of the undertaking, and tell us what was included and excluded: some thus explicitly record only those with timber value and exclude most or all of the pollards, whose short and often hollow boles rendered them of little interest to the timber merchant. Often, however, such limitations are not expressly stated, leading to possible errors of interpretation. Estate records of various kinds, including correspondence and accounts, can provide useful information about the management of farmland trees and woods, including details of the length of coppice and pollarding rotations and the age at which timber trees were felled, as well as informing us about attitudes to trees and the economics of forestry more generally. Unfortunately, much of the information they provide is impressionistic – little comprises the kinds of data that can be used statistically – and their allusions to tree health, in particular, are generally vague and hard to interpret. It is not always clear, moreover, what particular terms mean or, more precisely, whether they carry the same meaning in different contexts and different areas. *Dotterel*, for example, was usually employed as a term for a pollard, especially an old example, but in some cases simply meant any old tree. Maps are a particularly problematic source. Although there are a number of early examples which clearly attempt to show farmland trees with considerable care – sometimes even detailing the species and management of individual specimens – in many other cases it is unclear where, precisely, on the spectrum between accurate portrayal and mere illustration the representation of trees may lie. This problem is compounded by uncertainties about what, exactly, different sources mean by a 'tree', as opposed to a shrub or a sapling – a difficulty which, as we have already emphasised, bedevils all research into this subject. Some early surveys specifically record the number of young trees (often referred to as 'stands' or 'standils'), and distinguish them from mature trees and pollards, but many do not, leaving it unclear which ages or sizes were being excluded, or included, in particular documents.

Even apparently straightforward and relatively recent sources can cause difficulties. The first edition 6-inch and 25-inch Ordnance Survey maps from the later and (in parts of the North) mid-nineteenth century ostensibly show the location of every mature free-standing tree in the countryside, and this information has often been taken at face value by historians in the past, even though the original instructions given to the surveyors responsible do not appear to have survived (Rackham 1986, 222–37. Where comparisons can be made with near-contemporary estate maps they suggest that the numbers of trees depicted are in fact broadly, although not entirely, reliable. In south Yorkshire, for example, a selection of five estate maps produced around 1867 can be compared with the Ordnance Survey first edition 6-inch maps from the mid-1850s and the first edition 25-inch from the 1890s for the same areas.[1] In four cases the estate maps show tree numbers broadly in line with those depicted on whichever survey is closest in time. The exception, an estate map of a property in Ecclesall, shows significantly fewer trees, presumably because of differences in the size of 'mature' trees being included. Yet there are grounds for suggesting that in some contexts the depictions of trees on Ordnance Survey maps do need to be treated with more caution, not least because of the difficulties involved in showing trees where the map is otherwise crowded with features and symbols. A significant degree of omission thus seems to have occurred in settlements, where few trees are generally shown, but where other sources suggest they were often tightly clustered. Other possible problems include the extent to which hedgerow trees may have been omitted when tightly packed in hedges and, once again, how – and how consistently – the surveyors defined a 'tree', as opposed to a sapling or bush. This latter problem, which also affects our interpretation of such sources as the 1946 RAF vertical aerial photographs, makes comparisons over time of tree densities in particular areas very problematic. In short, few of the sources employed in this study were designed to answer the kinds of question which we are now asking of them, and this needs to be borne in mind in the pages that follow, even when not specifically highlighted in the text. This said, on the whole the available evidence provides a coherent story of the development of trees and woods over the last four centuries, if a highly complex and, at times, surprising one.

---

1   The relevant maps are: Ecclesall: Shf, WMP/MP/119; Swinton: Shf, WMP/MP/126; Brampton: Shf, WMP/MP/115 R; Hooton Roberts: Shf, WMP/MP/122; Nether Hoyland: Shf, WMP/MP/123 R.

 CHAPTER TWO

'One continued grove'

The density of farmland trees

How many trees are present in any area, how they are managed and how long they are allowed to live all have important implications not only for the appearance of the countryside but also for biodiversity: a landscape containing high densities of very old trees (for example) provides ideal conditions for certain kinds of wood-boring insect, while one in which trees of any kind are rare is inimical to most of our characteristic farmland birds. The ways in which rural tree populations were managed in the past may also have had an impact, as we shall see, on the general health of trees. For a large number of reasons, the management and density of trees in the countryside are thus of crucial importance, yet have received little attention from historians and others.

The most striking feature of the English landscape before the nineteenth century would, in many areas, have been the sheer density of trees. In the old-enclosed 'woodland' districts, especially in south-east England and East Anglia, but also in many western counties such as Shropshire, maps and surveys show that trees were densely crowded into hedges, far more densely than was to be the case by the later nineteenth century. In addition, they might be scattered across pastures or grouped into lines or bands two or three trees wide running around the field margins, often described as 'rows' or 'grovets' in early documents. A 1633 glebe terrier for the parish of Denton in Norfolk describes 'the Churchyard with a Pightell and a little Grovett above that Close towards the south', and 'Two Closes joineing together Westward called South Crofte – the first Close hath a Grovett above' (NRO PD 136/35). Such trees grew on the 'hedge greens' – strips of unploughed land around five or six metres wide on which cattle could be grazed and hay cut, but which also provided a place for the plough to turn while the field was being used as arable (Vancouver 1795, 87; Martin and Satchell 2008, 243–4). Some early maps also show single, double or occasionally triple lines of trees running through the middle of fields, probably where earlier boundaries have been removed. All this said, in most old-enclosed districts, and especially in those in which arable cultivation was of primary importance, the overwhelming majority of farmland trees grew in hedges.

While many early maps depict landscapes densely filled with trees, on only a few occasions did surveyors go further, carefully detailing the positions of individual examples and recording their species – and even in some cases whether they were managed as

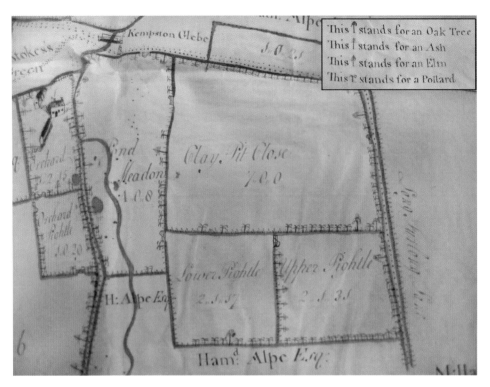

2.1 Section from a map of a farm at Beeston-next-Mileham in Norfolk, surveyed by Henry Keymer in 1764, with his key inserted. Note the phenomenal density of hedgerow trees. Those boundaries that apparently lack trees are simply the property of neighbours.

pollards or standards. Henry Keymer is one example. He prepared maps of a number of Norfolk estates in the middle of the eighteenth century in which he used different symbols to distinguish four main categories of hedgerow tree – elm, ash, oak and pollard (with that for a property in Scarning noting, in addition, the presence of alders) (Figure 2.1). Some of the farms he surveyed included a portion of open-field land – much of Norfolk had a mixed landscape – but on the enclosed portions the density of trees was high. A property at Beeston-next-Milesham, surveyed in 1764, comprised 70 acres (28 hectares), of which seven acres lay in the open fields and the remainder in hedged closes with an average size of 3.5 acres (1.4 hectares) (NRO WIS 138, 166X3). Keymer's map shows a total of 531 trees, an average of 7.6 per acre, or 8.4 per acre (c.21 per hectare) if we exclude the open-field land. An estate at Scarning, surveyed in 1761, was slightly larger, covering 80 acres (32 hectares), and lay entirely in enclosures with an average area of 4.2 acres (1.7 hectares) (NRO BCH 20). Keymer here shows a total of 1,095 trees, or 12.4 per acre (31 per hectare). A farm at Bintree, mapped in 1756, consisted of 79 acres (32 hectares), 15 of which lay in open fields and 64 in hedged closes with an average size of 3.4 acres (1.4 hectares), although this figure is slightly skewed by the inclusion of some very small yards around the principal farm (NRO Accn. Field 18.3.82 P188B). There were a total of 529 trees on the enclosed land, or

8.27 per acre (20 per hectare). These Norfolk properties surveyed by Keymer were not in any way unusual. An undated mid-eighteenth-century map of three farms in Whissonsett, for example, suggests a density of 21 trees per hectare (NRO HIL 3/34/1–27, 879X3).

All these places were located on the poorly draining boulder-clay soils that occupy most of central and southern Norfolk. The till in which they are formed continues as a low plateau southwards, beyond the river Waveney, forming the main central portion of Suffolk, and on into Essex and east Hertfordshire, becoming progressively more dissected and undulating, and the soils correspondingly lighter and better-drained. In all these areas the density of farmland trees could be even higher. A 'True Plan of the separate Pieces of Land in the Farm called Wises in the Parish of Widdington [Essex] with the number of uprights and pollards occupied by Mr James Perry', drawn up in 1756, lists 406 trees on a holding of just under 35 acres – a density of nearly 29 per hectare – but this figure is misleading, as much of the land on the farm still lay in open-field strips, without hedges, and the density on the *enclosed* portion was thus even higher (ERO D/DAb E2). A farm at Thorndon in Suffolk had as many as 29 trees per acre (*c.*72 per hectare) when surveyed in 1742 (Bury St Edmunds RO BT1/1/16), while on a property of 33 hectares at Badwell Ash in the same county a survey made in 1762 listed 115 timber trees and 650 pollards, but also noted that there were 'many hundreds more' of the latter (Bury St Edmunds RO E3/10/10 2/28; E3/10/10.2/32). But the prize for 'most densely timbered farm on the eastern claylands' must go to a 33-acre (*c.*13-hectare) property at Kelshall in north-east Hertfordshire, which had 24 trees per acre (59 per hectare) when surveyed in 1774 (HALS DE/Ha/B2112); however, an earlier survey of the same property, made in 1727, shows even higher densities, a staggering 99 trees per hectare, although this figure may include 'standils' or immature trees, in contrast to the documents just discussed, which seem to omit them (HALS DE/Ha/B2112).

For the most part, however, maps suggest an average of between 20 and 30 trees per hectare on the eastern boulder clays. For reasons that will become apparent, these figures are hard to compare with those of today, but they are significantly higher than those shown on the late nineteenth-century Ordnance Survey maps for the same areas. Of course, some surveys of properties on the eastern claylands suggest rather lower densities. Three large farms on the Tendring estate in south Suffolk had only 8, 12 and 14 trees per hectare in 1723, including pollards, the first of these figures relating to the home farm (Ipswich RO 1 HA 108/10/1). In some cases this was because not all the mature trees were valued – the surveyor excluding the pollards or, more rarely, some proportion of the timber. A terrier of part of Sewells Farm in Little Yeldham, Essex, drawn up in 1757, thus suggests a density of around 14 trees per hectare – 337 on a holding of 58 acres – but these were only the pollards, timber trees presumably being valued separately (ERO D/DAb E1). In other cases however a paucity of recorded trees was real and was the consequence of poor estate management and unreliable tenants. On the Flixton Hall estate in Suffolk in the 1720s it was said that one farm had been 'beggared by ploughing and stripped of nearly all its tree stock'. At Saxtead the tenant had taken down 'so many trees that there was not enough wood left for the dairy, so that firing had to be fetched from up to four miles away', while at Cotton Hall the tenant had 'managed to strip much of the farm of its trees and hedges' (Theobald 2000, 9–10). Overall, however, the densities of farmland trees recorded on the boulder clays of East Anglia, Essex and Hertfordshire were, before the mid-nineteenth century, extremely high by modern standards. The countryside was absolutely filled with trees.

The landscape of the East Anglian claylands was not unusual in this respect. In the far south of Hertfordshire and Essex, towards London, the soils formed in the London clay and other Eocene deposits were tenacious and difficult. Here, much of the land was laid to pasture, and many farms contained upwards of 25 trees per hectare. A particularly detailed map of Drakeshill Farm, Navestock, in south Essex, drawn up in 1722, records no fewer than 442 mature trees growing in the hedges and across the fields, more than 27 per hectare (ERO D/DU 583/2). On seven farms on the Broxbournebury estate in Hertfordshire in 1784, covering 373 acres (151 ha), there were 3,311 trees, an average of 22 per hectare: the greatest densities were recorded on West End Farm in Wormley, where there were no fewer than 1,506 pollards and 148 timber trees, together with 569 'spars' (saplings) and 18 apple and 2 walnut trees in the orchard, all growing on a mere 38 hectares – a density of around 40 mature trees per hectare (HALS DE/Bb/E27). At some places even higher densities were recorded, as on a farm at nearby Hammond Street, Cheshunt, in 1650, where there were there around 47 trees per hectare on the enclosed portion of the property (PRO/TNA E317/Herts/24, p. 37) Farms like this were, in effect, wood-pastures, subdivided by a mesh of hedges.

South of the Thames the densely treed landscape continued. The hedgerows of northern Surrey were 'very crooked and irregular, filled with elm and other timber', while in Sussex 'the enclosures are small' and 'each hedgerow is a nursery for timber' (Young 1808, 463). Here – as we shall see – many field boundaries, especially in the district known as the Weald, consisted of strips of woodland, comprising a wide band of coppiced understorey and variable quantities of timber trees, rather than hedges and hedgerow trees of normal form. But in the old-enclosed districts in the west – in Shropshire, Cheshire and Herefordshire especially – the hedges were like those in the east, and similarly thickly studded with trees. In Herefordshire John Clarke reported in 1799 that there was 'A great deal of very valuable timber … but it is so happily disposed of over the district, that it seems to scarcely interfere with the more substantial concerns of agriculture. Most of the timber is planted in hedge-rows' (Clarke 1794, 11). In Shropshire there were so many farmland trees that Plymley in 1803 complained that they blocked out extensive views across the landscape (Plymley 1803, 213); an eighteenth-century survey of land in the parish of Harley suggests a density of around 19 mature trees per hectare (SAR P100/F/1/1). The old-enclosed parts of north-west Dorset seem to have been particularly well filled with trees. An agreement concerning the felling of timber on land in Hermitage, drawn up in 1726, refers to 69 oaks and 50 ashes growing in a close called Twenty Acres and 51 oaks and

78 ashes in a close called Eight Acres (DRO D-1167A/4/1). In Warwickshire the 'numerous trees in the hedgerows of the old enclosures destroy the hedges … make the hedge sickly, and occasion gaps' (Murray 1813, 64).

The high densities of trees recorded in these old-enclosed districts in part reflected the small average size of fields and the tight mesh of hedges. Three small farms at Whissonsett in mid-Norfolk had more than 8.4 trees per acre, or 21 per hectare; but the fields (ignoring land lying in open fields) had an average area of around only 3.75 acres (1.5 hectares) (NRO HIL 3/34/1–27, 879X3). At Drakeshill Farm in Navestock, Essex, in 1722 there were 31 trees per hectare, as already noted, but the average field size, at under three acres (1.2 hectares), was miniscule (ERO D/DU 583/2). This said, there was no very close relationship between field size and tree density. Curd Hall Farm at Coggeshall in Essex had 16 mature trees per hectare in 1734 and an average field size of around 10 acres (c.4 hectares) ERO D/Dc E15/2); a property at Campsea Ashe in Suffolk had a slightly greater density (18 per hectare) when surveyed in 1807, but much smaller fields, averaging just 2.6 acres (just over a hectare) (Ipswich RO HD11:475). In part, this was because where fields were larger trees tended to be more densely clustered in the hedgerows. Overall densities could also be significantly raised by the presence of 'rows' of trees standing in from the margins of the fields, and timber or pollards more widely scattered across pastures. At Badwell Ash in west Suffolk in 1730 a 19-acre pasture field called Talbots Wood contained a total of 64 trees, while in 1762 18 timber trees were recorded there, along with 70 pollards growing 'out of the hedgerows', in addition to the 'many' growing within them. On the same farm in 1730 Bushey Mays, a pasture field of only five acres, contained 80 pollards growing 'all over' the field, while Priorland, covering 13 acres, had 'many pollards in the hedgerows and 60 pollards out of them' (Bury St Edmunds RO B E3/10/10.2/28). A tenant at Hindolveston in Norfolk in 1731 asked permission to fell 67 pollards in an 8-acre close (NRO DCN 59/21/1) – in effect, a small private wood-pasture.

Not all old-enclosed districts could boast quite these numbers of trees. North-east Norfolk was an area of relatively light, loamy soils, with large areas of heathland, and here tree densities were noticeably lower than on the nearby boulder clays. A survey made in 1722 of the estates of the Earle family, centred on Wood Dalling and Heydon, recorded 29,000 trees, of which c.24,000 were on farmland, mainly growing in hedges (NRO BUL 11/283, 617X2). Densities varied across the estate but ranged from around 7 to nearly 19 per acre (17 to 47 per hectare), averaging 11 per acre (27 per hectare); however, this figures includes, as well as mature timber trees and pollards, the 'storrells' or 'standrills': that is, small young trees. If these are omitted from the calculation there was an average of around 16 trees per hectare, significantly lower than in most of the places just discussed. A rather smaller survey of property at nearby Burgh and Skeyton, dating from 1816, suggests similar densities (NRO BR 90/35/22).

The Chiltern Hills – lying to the east of the boulder clays in Hertfordshire, and extending west into south Buckinghamshire and Oxfordshire – were in many ways similar.

They were characterised by relatively light and acidic clays, largely enclosed from an early date, and included extensive tracts of wooded common. On three farms lying in Hemel Hempstead, Flamstead and Redbourne and recorded in a survey of 1838 there were fewer than five trees per hectare, while a map and survey of 'Three Fields called Church Lands' in Watford, made as early as 1754, recorded fewer than three per hectare (HALS DE Cr 55; HALS DP 117 25/7). On the other hand, an undated late seventeenth-century map of a 49-acre (*c*.20-hectare) farm in Flaunden, close to the Hertfordshire/Buckinghamshire border, suggests a density of over 15 per hectare, so there was evidently much variation (HALS DE/X905/P1). All this said, most of the aptly named 'woodland' districts in the south and east, as in the west, boasted phenomenal numbers of trees. Humphry Repton, writing at the end of the eighteenth century about north Norfolk, described how the hedgerow trees were so numerous that they gave 'a prodigious softness to the landscape, that in many parts appears to be one continued grove of many miles extent' (Armstrong 1781, III, 3). It was a comment he might have made of most of the old-enclosed parts of England.

The abundance of pollards

An equally striking feature of the countryside before the mid-nineteenth century in most of these old-enclosed districts was the phenomenally high proportion of farmland trees that were managed as pollards. As already emphasised, these are not always fully visible in our sources. Some explicitly omit them: a survey of Tarrants Farm in St Stephens, Hertfordshire, made in 1733, for example, simply states 'Pollard trees not reckoned', while a valuation of timber at Little Cowarne in Herefordshire in 1811 noted that 'In addition to the above there are many pollards not brought into this return and also a quantity of small oak and ash' (HAR F55/21/23). Others omit them silently, and perhaps partially – noting only those which, if felled, might have had some timber value. Where maps or surveys are comprehensive and reliable, however, they usually reveal that upwards of three-quarters of the trees present on particular properties were pollarded, and sometimes far more.

We might start, once again, with the long-enclosed and bosky countryside found on the heavy boulder clays of East Anglia, Essex and Hertfordshire. On the estates mapped by Henry Keymer at Beeston and Scarning in Norfolk, for example, 79 per cent and 76 per cent of the trees, respectively, were pollards, while his map of a farm in Wendling in 1777 shows that 73 per cent were so managed (NRO EVL 348/1/1 and 29). A survey of 'Mr Fromours Estate' in Pulham in the same county, made in 1757, records 41 timber trees but no fewer than 472 pollards: even if we count the additional 46 'stands', or young trees, with the timber, the pollards still made up over 84 per cent of the trees on the farm (NRO DN/MSC 4/22–25). In 1803, on a farm at Loddon, 76 per cent of the recorded trees were pollards (NRO MC 78/15, 522x7), while a timber survey of 1813 suggests a figure of 89 per cent (excluding saplings) on a farm at Kenninghall (NRO HNR 149/3). The situation was no different further south, where the boulder clays continued into Suffolk and Essex. On the 35-acre farm in Widdington in the west of Essex in 1756 there were 322 pollards but only 84 timber trees – that is, 80 per cent of the trees were pollards (ERO D/DAb E2); on the three farms at Tendring surveyed in 1733 the figure was likewise 80 per cent (Ipswich RO 1 HA 108/10/1); while on a property at Chevington in Suffolk in 1820 79 per cent of the 264 trees recorded were so described (Bury St Edmunds RO HA 507/2/460). Occasionally, the proportion was even higher, rising above 80 per cent. At Thorndon in Suffolk in 1742 82 per cent of the trees were pollards (Bury St Edmunds RO BT1/1/16); on a farm in Finchingfield in Essex in 1805 the figure was 83 per cent (ERO D/D Pg T8); while on a farm at Campsea Ashe in Suffolk in 1807 pollards constituted no less than 94 per cent of the trees (Ipswich RO HD11:475). At Curd Hall Farm, Little Coggeshall, in 1734 there were no timber trees at all of ash, oak and elm, although there were 688 'spire', or young trees, of uncertain species, together with 10 mature hornbeam and 36 walnut. There were, however, no fewer than 3,591 pollards (ERO D/Dc E15/2). As we shall see, the numbers of pollards declined from the late eighteenth century, but the process was a gradual one. On Thurston Hall Farm, in Hawkedon, Suffolk, there were as late as 1838 683 pollards out of a total of 1,001 mature trees (Bury St Edmunds RO HA535/5/35). Stray comments in letters, journals and valuations similarly imply that pollards were usually much more numerous than timber trees. Typical was the comment made in 1792 by the attorney and maltster Meadows Taylor, who described how there were 'several good timber trees and many pollards growing upon the lands' at Palgrave in Suffolk (NRO MC 257/23/5, 638 x 8).

Where the boulder clays extended into east Hertfordshire similar numbers of pollards could be found. At Barwick in Standon in 1778, for example, 85 per cent of the trees on three farms and 84 per cent of those on the manorial demesne were pollarded (HALS A/2832 and 2833). Pollards accounted for 83 per cent of the trees on land at Whempstead, Little Munden, in 1808, and the same on Pearces Farm in Thorley and Sawbridgeworth in 1807. On the Cowper estate around Hertingfordbury 91 per cent of the trees recorded on a map of 1704 appear to have been pollards (HALS D/EP/P4), while the 1727 survey of Kelshall, already mentioned, reveals not only a phenomenally high density of trees but also that a very high proportion – no less than 93 per cent – were pollards (HALS DE/Ha/B2112). Even in 1774, when a further survey was made of this 33-acre property and the numbers of trees had been significantly reduced, 86 per cent were still being managed in this manner (HALS DE/Ha/B2112). Although, as we shall see, the vast majority of the pollards that once filled the landscape of these districts have since been removed, a small proportion remain and ancient examples crammed tightly along field boundaries can sometimes be found, often hidden away in remote locations – vestigial traces of these lost, bosky landscapes (Figure 2.2).

On the poor soils immediately to the north of London the proportion of pollards recorded by seventeenth- and eighteenth-century surveyors and commentators was likewise very high. In Middlesex, Middleton noted in 1798 that 'many of the hedge-rows of this county are disfigured by pollard trees', and described how the hedgerows 'almost

2.2 Ancient oak pollards in a hedge in south Norfolk, survivors from a large and dense population. The existence, in many places, of magnificent veteran trees like these does not necessarily indicate that the countryside was once filled with such ancient specimens. In the past, most pollards were probably replaced long before they reached this kind of age: these particular examples were, perhaps, little more than two centuries old when active management ceased.

every where disgust us with the sight of rotten pollards' (Middleton 1798, 275, 278). In 1630 there were 709 pollards but only 252 timber trees on the demesne of Boreham manor in south Essex, together with 11 'saplins', or young trees: that is, pollards accounted for 74 per cent of the mature trees (ERO T/A 783/1), as they did on a farm in Hatfield and North Mymms in Hertfordshire surveyed in 1723 (Austin 2013). But at Navestock in Essex in 1722 pollards made up no less than 91 per cent of the mature trees on the farm (ERO ERO D/DU 583/2), while on seven farms on the Broxbournebury estate in south Hertfordshire in 1784, covering 373 acres (151 ha), there were 3,012 pollards but a mere 299 timber trees (HALS DE/Bb/E27): the greatest density here was recorded on West End Farm in Wormley, where there were no fewer than 1,358 pollards but only 148 timber trees. On the Broxbourne Mill estate in the late eighteenth century over 80 per cent of the trees on the higher ground seem to have been pollards, and on the meadows the figure may have been as high as 93 per cent (HALS B479), while an 89-acre farm at Hammond Street, Cheshunt in 1650 apparently contained no timber trees at all, but as many as '927 lopt Pollardes of oake Ash and Hornebeame' (PRO/TNA E317/Herts/24, p. 37).

Not every property in these old-enclosed 'woodland' districts lying to the south-east of the 'champion' regions could boast quite these numbers. On Sir John Henniker's estate in Waltham Abbey in south Essex, for example, in 1791 pollards made up only 64 per cent of the trees recorded outside the woods and copses (1,801 out of 2,812) (Ipswich RO HA 116/5/11/2). Some extensive districts within this wider 'woodland' region – generally those in which the density of farmland trees was itself less – likewise displayed noticeably lower numbers, such as the light loams of north-east Norfolk. On the Earle estates in 1722 only around 64 per cent of the trees were pollards (NRO BUL 11/283, 617X2); on a property at nearby Burgh and Skeyton in 1816 the figure was 63 per cent (NRO BR 90/35/22); while on an estate at Gunthorpe in 1775 only 37 per cent were managed in this way, although the survey may exclude pollards with no timber value (NRO BL/CS 1/20/3/1–5). In the Chiltern Hills in Oxfordshire, Buckinghamshire and west Hertfordshire pollard numbers were also low on many farms, while, as we have seen, in the Weald of Sussex and Kent fields were often enclosed not by hedges of normal form but by 'shaws', narrow linear woods consisting of standard trees growing among coppice: Young described 'corn surrounded by a forest in every hedgerow' (Young 1808, 180). In such situations, poles could be obtained in large quantities from the coppiced underwood and, while some pollards existed, they were fewer in number. But in most of the 'woodland' districts of the Home Counties and East Anglia at least 70 per cent of farmland trees were pollarded, and often 80 or even 90 per cent. The countryside would thus have looked very different from today, with most trees either recently shorn of their branches or in various states of regeneration: a mass of lollipop-shaped features, never attaining the tall, spreading forms we are used to seeing.

The situation was much the same in the old-enclosed 'woodland' districts lying to the west of the 'champion'. In Herefordshire Clarke noted in 1794 that hedgerow trees were mostly pollarded (Clarke 1794, 11, 67); a little earlier it was reported that 'The Trees are much strip't and lopp'd by the Farmers', to such an extent that the only good timber was to be found in woodland (*Journal of the House of Commons* 1792, 318). In Shropshire farmland trees were 'generally found most decayed in consequence of lopping' (Plymley 1803, 213); 'tenants have no scruples about lopping the best trees for fuel, or other purposes, provided it can be done with impunity' (Plymley 1803, 216); and 'the Occupiers of Lands frequently abuse such [trees] as grow in Hedge Rows' (*Journal of the House of Commons* 1792, 319). This said, there was much local variation and, in general perhaps, the sources hint at lower densities of cropped trees than were to be found in the south-east. A survey made in 1690 of lands in West Felton in Shropshire suggests that only around a quarter of the trees were pollards (SAR D 3651/B/4/2/238/21). On the Clehonger estate in Herefordshire in 1820 37 per cent of the trees recorded were pollards (HAR C38/49/3/1), while on an estate in Bircher in Herefordshire in 1828 pollards made up only around a fifth of the trees recorded (HAR F76/111/50). In these surveys, as in others that survive from this region, it is possible that only pollards with a timber value were being recorded. But, against this, we should note sporadic evidence that landowners here were, by the eighteenth century, apparently making more strenuous attempts to limit pollarding than was the case in the south-east or East Anglia. A lease from Kynaston in Shropshire, from 1787, thus instructed the tenant not to lop trees 'except such Runnels as have usually been cropped for repairing fences on the premises only' (SA 867/25); while leases for land at Bishop's Castle, from 1728, and Ludford, from as early as 1683, contained clauses prohibiting the pollarding of potential timber trees (SA 11/537; SA 11/535). Nevertheless, pollards were clearly a common if not dominant feature of the landscape of these western 'woodland' districts. Benjamin Slade in 1771 thought that 'Hereford and Shropshire could produce a great quantity of timber if the people would let it stand but it is cut too soon so will only produce pollards' (PRO/TNA ADM 106/1205/58). In the old-enclosed parts of Gloucestershire the hedgerow oaks were 'usually pollarded and lopped' (Rudge 1807, 242), while in south-east Somerset there was 'plenty of wood in the hedges, and on the pollard trees, but few timber' (Billingsley 1794).

In all these old-enclosed districts, both east and west of the 'champion', references in leases and other documents suggest that pollards were usually cropped at the same time as the hedge in which they grew was cut back and the associated ditch dug out, usually at intervals of around ten years. But both hedge maintenance and the lopping of pollards could occur on a shorter rotation. In the late eighteenth century copyhold tenants of the manor of Barnet, close to London, were instructed 'Not to Plash any hedge or Lop any Pollard, till it has attained at least seven years growth, and then to scour and cleanse the ditch at the same time' (HALS DE/B983/E1). Only in a minority of cases did leases stipulate the frequency with which hedges should be 'new made'; but, where they did not, the frequency with which pollards should be lopped might nevertheless be stipulated. A lease from Copdock in Suffolk drawn up in 1775 ordered the tenant to 'keep pollard trees under a regular fall of 9 years growth under penalty of 9 shillings per tree' (Rackham 1986, 220).

Trees in the 'champion'

Between the old-enclosed areas of south-east England and the Home Counties and the well-hedged lands of Herefordshire, Shropshire and the neighbouring counties lay the 'champion' landscapes, in which, as we have seen, extensive tracts of open-field survived well into the eighteenth century and in some cases into the nineteenth. It is generally assumed that in this region most parishes contained few hedges, and thus few hedgerow trees, until they were enclosed. Indeed, early advocates of enclosure made much of the paucity of wood and timber in open-field districts, and the hardships that this engendered. In reality, the situation is more complicated. To begin with, we should note that in all such areas a network of enclosed land – hedged (or walled) tofts and crofts – existed in the immediate vicinity of the nucleated villages, and these often appear to have been very densely planted with trees. Early drawings and sketches – such as those of the Northamptonshire landscape prepared by Peter Tillemans in the 1720s – show distant villages almost as small woods (Bailey 1996). Documentary evidence supports this impression. In 1377 at Ravensthorpe in Northamptonshire a house with three tie beams was built entirely using timber growing on the toft belonging to the messuage (NHRO LT64). A detailed survey of the manor of Milcombe in Oxfordshire, made in 1656, includes a list of 'what wood and timber is growing on the premises' of the tenants. The hedges of the village closes were liberally stuffed with trees. To the rear of Thomas Burchall's farmhouse, for example, the surveyor recorded '70 trees of Ash and Elm 18 of them be very small samplars and there be 48 willows and 2 maple trees'; while in the 'close and orchard' behind Thomas Strainke's farmstead there were '126 trees of Ash and Elm whereof 30 of them be very small samplars, in these 2 places there be 52 withes [willows] small and great and about as many new planted' (NHRO C(A) Box104 4 1656). The trees in tofts and crofts were, moreover, constantly being augmented. A lease drawn up in 1582 for land in Crick in Northamptonshire, a township still entirely unenclosed, stipulated that the lessor was to plant five ash, oak or elm trees every year, presumably in the toft behind his farm (NHRO A/089). Planting was not confined to the messuages of tenants, however, but might embrace more public areas of settlements. Morton, writing in 1712 about Northamptonshire, described the numerous large elm trees growing 'in or near the streets of many of our small towns upon a green bank or some other convenient By Places' (Morton 1712, 396).

What is more surprising is that the open fields lying between the villages were by no means devoid of trees, or of hedges. The latter often existed on the boundaries between adjacent parishes, even when both remained otherwise unenclosed. A document of 1735 thus describes the making of a new hedge between Wilby and Wellingborough in Northamptonshire, parishes not enclosed until 1801 and 1765 respectively, detailing payments not only for hedging plants but also for willows, presumably intended for trees (Hall 1995, 264; NHRO 350P/90). Hedges, which we can assume usually contained trees, also sometimes surrounded the individual open fields (Hall 1995, 244), the parish meadows, areas of common land and 'cow pastures' – that is, parcels of land taken out of the arable in the post-medieval period to increase the area of common grazing. In addition, early maps sometimes show discontinuous fragments of hedge lying within the open fields themselves. Good examples can be seen on maps of Northamptonshire parishes such as Murcott (1771) (BL Add. MS 78141 A), Wollaston (1789) (NHRO 447), Brigstock (1734) and Broughton (1728) (Boughton House archives). An undated document relating to Ecton in the same county suggests their possible significance, referring to 'arbours and shelter-hedges in the beasts and sheep pastures' which were to be 'preserved and kept for the sheltering of herdsman and cattle as according to the ancient and laudable manner'. The responsibility for their maintenance lay with the lord of the manor, who was to have 'the benefit of the lopping … and plashing of the said arbours and hedges … ' (NHRO E(S) Box X1071).

In addition to trees present in hedges, free-standing examples could often be found on commons and cow pastures, to judge from the evidence of seventeenth- and eighteenth-century maps, as in the Northamptonshire parishes of Weekley (NHRO 1349), Quinton (NHRO 2895), Easton-on-the-Hill (NHRO 6-p/504) and Wollaston (NHRO 4447). Examples are also sometimes depicted lining roads, usually major ones, running through the open fields. The enclosure map for Ecton, drawn up in 1759, shows trees growing along both sides of the road leading from Northampton to Wellingborough. The enclosure award, in confirming the course of the road and establishing its dimensions within the new enclosed landscape, specifically states that it should extend 'three feet in breadth as well on the north as on the south side of the elm trees now standing or growing theron and belonging to Ambrose Isted [the lord of the manor] so as to include the said trees into the said road' (NHRO Inclosure Vol A p9; Partida 2014). Trees are even occasionally shown on early maps growing within the arable furlongs, as on an earlier map of this same parish, surveyed in 1703 (Partida 2014, 166; NHRO Map 2115); a map of Kirby, surveyed in 1580 (NHRO FH 27285); and a map of Woodford, surveyed in 1731 (Boughton House archives). As Partida has noted, crab apples feature regularly in the names of furlongs, as at Cogenhoe, Aynho, Weekley, Cranford St John and Higham Ferrers (Partida 2014, 165–8): a map of Ecton, surveyed in 1703, shows a probable example (Figure 2.3). With its twisted and knotty grain, crab apple was employed for a number of specialist purposes, including the construction of cog wheels in wind and water mills, and its relatively small size perhaps made it ideal for planting on the ends of arable strips. Trees growing on the commons and wastes were in most cases the property of the manorial lord, but those growing on the ends of arable strips presumably belonged to the strips' proprietors. In a legal case concerning Kettering in Northamptonshire in 1807 one witness described how his uncle, Thomas Collis:

> Was owner of some copyhold lands in the Open fields of Kettering near the Turnpike Gate on the Road to Barton Seagrave the adjoining Parish. On which Lands he T Collis in the winter about or between the years 1750 and 1760 cut down a very large and high ash tree and also a large oak tree (NHRO GK10).

2.3 Detail from a map of Ecton, Northamptonshire, surveyed in 1703, showing a lone tree (probably a crab apple) growing among the furlongs.

The greatest concentrations of trees in many Midland open-field parishes were, however, to be found on the flood plains and meadows. This is clear not only from early topographic views, such as those of the Northamptonshire countryside prepared by Tillemans, but also from maps and in particular from the claims made at the time of enclosure, when compensation was paid to the owners of particular trees that were now incorporated within someone else's private land. The Northamptonshire parish of Irthlingborough still lay almost entirely open when it was enclosed by parliamentary act in 1808: the list of claims drawn up records a total of 3,055 trees, a density of two per hectare, averaged across the parish (NHRO ZA 906). No less than 62 per cent were willows, presumably growing on the flood plain of the river Nene, which ran through the parish. Given the purpose of the document, this figure almost certainly excludes most of the trees in the village 'envelope', which were unaffected by enclosure or exchanges. A similar valuation was drawn up on the eve of the enclosure of Finedon in the same county in 1806. This recorded 1,038 trees (although 55 were 'thorns'), a density across the parish as a whole of around 0.7 per hectare, but which must again exclude most of the trees packed into the 'ancient enclosures' around the village itself (NHRO ZA898). Again, the large numbers of willows (410), presumably growing in the meadows beside the river Ise, is noteworthy. Another such valuation, for Wilby in 1802, lists 433 trees within a parish extending over 466 hectares – just under one per hectare, again including numerous willows. An earlier valuation of timber in the parish, made in 1764, lists 762 trees – an average of 1.6 per hectare across the parish as a whole, nearly a third of which were willows (NHRO X1657). In neither case is it entirely clear what proportion of these were growing within the closes around the village, but the nature of the former document and the abundance of willows suggests that most stood out in the fields and meadows, some perhaps in the hedge on the boundary with Wellingborough, which, as noted earlier, had been planted with willows in 1735 (NHRO 350P/90).

Other documentary sources suggest similar overall tree densities within unenclosed parishes. A survey of 1749, listing the timber growing on the lands of the Boughton estate in Northamptonshire, details 517 trees in the still unenclosed parish of Luttington, around 80 per cent of which was in the hands of the estate, implying a density of around 1.2 per hectare. Other evidence, however, makes it clear that much of the property within the village envelope itself remained in the hands of small proprietors, suggesting that this figure again needs to be adjusted upwards (Boughton House archives). The same source records 891 trees in the largely open parish of Geddington, where the estate owned around half the 678 acres, perhaps suggesting a density of around 2.6 per hectare across the parish as a whole.

We noted earlier that there was much variation in the character of open-field landscapes, and most of the examples so far discussed come from areas of heavy soil. Yet where 'champion' countryside was found on light land – on chalk or sand – the situation was broadly similar. On the light soils of west Norfolk, formed in chalk and sandy drift, a map of the open-field parish of West Lexham, surveyed in the late sixteenth century, shows large numbers of trees in village closes and growing in demesne enclosures around the manor house, together with examples scattered across some of the village commons and a particularly dense concentration (presumably comprising alders or willows) on the banks of the river Nar (NRO MS 21128, 179X4). Even though it still lay almost entirely unenclosed, more than 1,230 trees are shown, a number which – if roughly accurate – would suggests a density, averaged across the parish as a whole, of around 2.6 per hectare.

It is important to emphasise the numbers of trees in 'champion' districts before enclosure because most accounts present a rather different picture, of landscapes comprising endless, uninterrupted vistas of ploughland. As in old-enclosed districts, moreover, many of the trees were managed as pollards. Over half of the trees growing in 1749 on the Montagu estate lands in the Northamptonshire parish of Geddington, which lay almost completely unenclosed until 1807, were pollards; on their property in Hanging Houghton, likewise entirely open at the time, the figure was 78 per cent. A survey of Great Doddington made in 1764, when the parish still lay open, lists 380 trees growing on one estate, of which 200 (53 per cent) were dotterel (pollarded) ash and a further 85 (20 per cent) dotterel 'salley', or sallow. Pollards accounted for 100 per cent of the trees recorded on land in the open-field parish of Ickburgh in west Norfolk in 1651 (NRO WLS XXVI/4, 414X6); around 56 per cent of those on a farm at Hockwold in 1816, in the year that the parish was enclosed (NRO BR 90/34/16); 75 per cent of the trees recorded on a property at nearby Mundford in 1805 (a parish enclosed by an act passed in the following year) (NRO MS 13751, 40E3); and a similar proportion of those valued in the exchanges when Griston, some 15 kilometres to the east, was enclosed in 1777 (NRO C/Sca 2/135).

Although in many parishes, as we have seen, the open fields remained intact and flourishing into the later eighteenth or even nineteenth centuries, in others they did not. Some places were enclosed in their entirety well before parliamentary enclosure began in earnest in the 1750s and 1760s. In others partial enclosure occurred, usually as owners

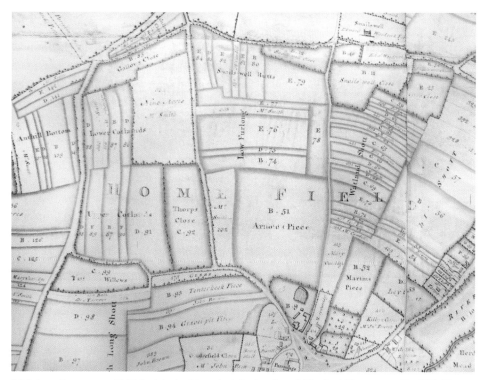

2.4 Extract from a map of Ickleford, in the 'champion' district of north Hertfordshire, surveyed in 1771. The open fields have been partly enclosed through piecemeal exchange and consolidation. The hedges bounding these enclosures are studded with trees: other can be seen lining the roads, scattered across meadows and closely clustered in the village closes.

sold and exchanged strips in a piecemeal fashion, creating patches of hedged ground within the otherwise open arable. The parish of Flore, to the west of Northampton, for example, still lay mainly open in 1704, but had undergone some piecemeal enclosure, and a terrier drawn up in that year describes '1 Yardland of arable, meadow, and pasture ground within the open and common fields of Floore'. The property included one headland in *Offerlong* 'with the hedge'; land in Middle Field that included grass and meadow ground with three willows; seven willows in 'Mernes' and four in Church Hooke; the hedge at 'outland' and another hedge on Collins Hill; and land lying 'west of … Collins Hill and the hedge and trees in the hedge' (NHRO ZB0640/007). Similarly, a map of Ickleford in north Hertfordshire, surveyed in 1771, seems to show (albeit schematically) large numbers of trees growing on flood plains, on roadsides and commons and in closes around the village, as well as in partial enclosures within the fields – perhaps as many as five per hectare, averaged across the whole parish (Figure 2.4) (HALS DE/Ha/P1).

Piecemeal enclosure probably always increased the number of trees. But where open fields were removed entirely, at a stroke, the results may have been more variable. The earliest such enclosures in Midland parishes – those occurring before the middle of the seventeenth century – were often imposed by large landowners and involved a shift of land use from arable to permanent pasture, a development often associated with the depopulation and destruction of the village itself. The sheep were grazed in large fields, occasionally as large as the old open fields had been (Taylor 1975, 115–17). The Northamptonshire village of Papley, for example, was depopulated and laid to grass in the 1490s, and a map surveyed in 1632 shows an average field size of nearly 20 hectares (NHRO Map 2221). Given that depopulating enclosures like this led to the neglect and eventual destruction of the tight meshes of closes around the village itself, there was often little overall increase in the length of hedges, and sometimes even a reduction, potentially leading to a fall, over time, in the numbers of farmland trees. On the other hand, in many cases the pasture fields might themselves be scattered, to varying extents, with free-standing trees, as indeed was the case of Papley, although the admittedly schematic map suggests that their numbers may have been limited: only around 280 were shown in the whole parish, less than one per hectare. From the seventeenth century, however, large-scale formal enclosures do seem to have led to a significant proliferation of trees. Rather than being imposed from above, enclosure was now normally achieved through agreement between the principal proprietors, and was less likely to be associated with village shrinkage and depopulation. The landscape following enclosure was often used for mixed farming rather than large-scale sheep ranching. Not only were the fields smaller, but the villages survived, together with the densely treed enclosures around them. The overall length of hedges, and thus the numbers of hedgerow trees, increased as a result, although they seldom reached the levels found in anciently enclosed districts in the south-east or west of the country.

This said, the evidence of both maps and surveys suggests much variation. In some parishes the numbers of trees remained surprisingly low. Nobottle in Northamptonshire was enclosed by agreement towards the end of the seventeenth century and a map of 1715, which seems to plot trees with some care, suggests a density of around 4.2 per hectare (BL A DDMSS 78143). Surveys of the timber on the Boughton estate in the same county, made in 1749, indicate that the two Barnwells, both enclosed in 1683, had together around three mature trees per hectare; Hemington, enclosed around 1657, had 3.3. These figures were exceptional, however. Armston on the same estate, enclosed in 1683, had five per hectare and Kingsthorpe no fewer than six (Boughton House archives), while figures more than double this were known from elsewhere in the county. A survey of enclosed land on the Duke of Powis' estate in Upper and Nether Heyford, Glassthorpe and Newbold made in 1758, covering some 770 acres, records a total of 3,004 trees, or 9.6 per hectare, although even this seems to exclude willows, as the surveyor made comments such as '29 Ash and Elm and also many willows' (NHRO ZB 1837). On John Darker's estate in Gayton, Tiffield, Kislingbury, Milton, Litchborough and Upper Heyford in 1791 no fewer than 10.6 trees per hectare were recorded (NHRO YZ 2183). Even these figures, however – usually in the range of 5–10 trees per hectare – were significantly lower than those found in old-enclosed districts, where 20, 30 or more were found in places such as Hertfordshire and Norfolk.

What is also striking is that, as enclosure progressed through the later seventeenth and eighteenth centuries, the proportion of trees managed as pollards in these districts seems to have declined. In nine townships on the Montagu estate in 1749, all enclosed during the previous half-century or so, between 9.5 and 25 per cent of the trees were pollarded. In Oakley and Brigstock, parishes that were not fully enclosed until 1807 and 1795 respectively but where the majority of the estate land lay in enclosures made piecemeal from the fields, 22.8 per cent of the recorded trees were pollards. Similarly, 26 per cent of the trees listed on John Darker's estate in 1791 were described as pollards (NHRO YZ 2183). The contrast with the proportions apparently present in the pre-enclosure landscape, or existing contemporaneously in the old-enclosed districts – where upwards of 70 per cent of trees were routinely cropped – is striking.

Trees in the north

In the far north of the 'champion' belt, in the open-field countryside of the Vale of York, the Wolds and Holderness, there seem to have been even fewer farmland trees, even after enclosure. Maps of enclosed properties in Yorkshire parishes such as East Harlsey (1762), Pickhill (1778), Birdforth (1770), Easby (1779), Tadcaster (1718) and Helmsley (1785) – all of which appear to mark the positions of trees with some care – show densities of between 0.4 and 2.5 trees per hectare, averaging 1.4 per hectare (Nrth ZNS; Nrth ZIQ; Nrth ZDS M 2/12; Nrth ZMI; Lds WYL68/63). A map showing the whole of Warmsworth parish, then partly enclosed, suggests an overall density of 1.8 per hectare, rising to 4.3 per hectare within the enclosed land (Don DD/BW/E11/7). Willow pollards may have been a common feature in some Vale townships, as they were in the 'champion' districts of Northamptonshire, and examples still survive in places such as Fishlake, to the north-east of Doncaster (Jones 2012,

141). But, on the whole, trees remained rare in the champion Vales. In old-enclosed areas of the north, such as the Lancashire lowlands or the Dales cutting through the Pennines, the situation was rather different. Some evidence suggests that, in the seventeenth century, these districts might have boasted a significant number of farmland trees, although never as many as the old-enclosed areas in the south-east of the country. Richard Muir has studied a group of 186 ancient oak pollards in Nidderdale, the overwhelming majority of which either grow, or originally grew, in hedges (Muir 2000, 100). In some cases, individual examples can be correlated with trees shown on early nineteenth-century maps, and seem to be survivors of a denser population, perhaps reaching 15 per hectare at places such as Clapham. Away from champion 'vales', in the Pennine Dales or on the margins of the uplands, farmland trees may thus once have existed in some numbers at the start of the period studied here. In most districts, however, numbers seem to have declined through the seventeenth and eighteenth centuries, and maps of old-enclosed Yorkshire townships such as Walburn (undated, early eighteenth century), Leyburn (*c.*1730) and Downholme (1738) – all located in or on the margins of the uplands – show densities ranging from 1 to 2.7 per hectare, averaging around 1.5 (Nrth ZAZ (M) 1; Nrth ZBO (M) 1/2; Nrth ZBO (M) 1/5). By the end of the eighteenth century, while there was said to be a 'considerable quantity' of timber growing in the hedgerows of Lancashire, on the whole 'sun-shine is preferred to shade' (Holt 1794, 61). Tree numbers, already lower than in the south, probably fell in many northern areas through the seventeenth and eighteenth centuries.

The available evidence also suggests that the proportion of farmland trees managed as pollards also declined through the seventeenth and eighteenth centuries. As noted, Muir has pointed to the prominence of pollards in the boundaries of Nidderdale in the seventeenth century, and some of the early pollards noted by Fleming in nearby Swaledale were likewise boundary trees, rather than relics of early wood-pastures (Fleming 1998). In Cumberland, Thomas Pennant drew attention to the numerous ash pollards that were cut for fodder (Pennant 1776), and in places the remains of these populations can still be seen, growing against or, more rarely, within stone walls, especially in Borrowdale and Langdale (Figure 2.5). When, following the Dissolution, property belonging to Roche Abbey at Barnby in south Yorkshire was sold, the property included oaks and ashes that were 'moste parte usually cropped and shred' for 'housebote' (building repairs) and 'hedgebote' (hedging stakes and materials) (Aveling 1870, 102). Leases and accounts from various northern districts also refer to the right of 'greenhew', normally a reference to pollarding – perhaps for fodder – although in some cases it could be exercised only to a limited extent, as at Settrington in Yorkshire, where the tenants could only cut 'small writhing [bendy] bands for tying up ther cattell & making Harrow withes' (King and Harris 1963, 10). All this said, by the middle of the seventeenth century pollarding was already in decline in some districts, and rare in many by the eighteenth. A survey made in 1642 of trees at Helmsley, on the northern margins of the Vale of Pickering in Yorkshire, shows that the 'dotterels' made up only 14 per cent of the trees recorded (1,269

2.5 Borrowdale, Cumbria: ancient ash pollards growing next to drystone walls. In the Lake District, as in some other districts of northern England, the practice of using ash or holly as winter feed for livestock persisted into the eighteenth century.

out of 9,234) (Nrth ZEW IV 1/6). Eighteenth-century surveys and other documents from most northern areas seldom refer to pollards on farmland at all and, when they do so, usually appear to describe *former* pollards, such as the 'distorted rickety trees of elm and ash' noted in late eighteenth-century correspondence from the Bolton estate, or the '166 oakes or hedgehogges' noted in a valuation made on the same estate in 1722 (Cald H1/ YT/1772/Jan 15). Tuke in 1800 could still describe how the tenants 'on some estates' in Yorkshire were allowed to crop hedgerow trees, 'in doing which, large boughs are frequently cut off in a very rough manner' (Tuke 1800, 191). But he implies that this was relatively unusual, and it is striking that agricultural writers in the decades either side of 1800 seldom mention pollards in the north in the way that they so often do in the old-enclosed countryside of the south-east, East Anglia and the West Midlands. In short, even in the seventeenth century tree densities in most parts of northern England were probably lower than in the south, and they declined steadily thereafter; by the second half of the eighteenth century there were relatively few farmland trees in most districts, and fewer still that were actively managed as pollards. This is mirrored, to some extent, in the present landscape. While ancient pollards can be found in the north which evidently grew in enclosed landscapes, as in Borrowdale or Nidderdale, most are associated with former wood-pastures, and especially deer parks – such as the large and picturesque collection in the park at Chatsworth, in the southern Pennines.

The management of farmland timber

The very high numbers of pollards is the most surprising characteristic of early tree populations in most areas of England. But perhaps equally striking is the young age of most timber trees. The majority of our documents describe the size of trees in terms of the volume of wood they contained, although it is not always clear whether these are given in cubic feet or in Hoppus feet (developed in 1736 by Edward Hoppus, a unit to take account of the losses incurred when a log or stem is converted into timber: a Hoppus foot is 1.273 times the size of a cubic foot. A few documents list the quarter-girths of trees, a measurement which, when squared, provides a good approximation of the area of the square that can be fitted within the circumference of a tree, which is itself a good prediction of the volume of usable timber. It is not easy to estimate the age of trees from their volumes: quarter-girths provide better evidence because, as dendrologists have long suggested, the circumference of a tree has a closer relationship to its age than any other dimension (White 1998). Most oak and elm trees growing in relatively open situations will attain a girth of a metre within about 40 years and a girth of two metres in around 80 years; ash trees slightly less. Accounts from an estate at Aspall, just north of Debenham in Suffolk, dating to the period between 1728 and 1763, are unusual in describing some trees in terms of both volume *and* circumference (private archives). They suggest that farmland oaks with quarter-girths of around eight or nine inches (around a metre of full girth) contained 20–25 cubic feet of timber, but sometimes as much as 35 (private archive); while those with quarter-girths of around 18 inches (around two metres of full girth) contained around 40–50 cubic feet. These equivalences are, broadly speaking, confirmed by discussions with modern woodsmen and foresters. In other words, hedgerow trees that our early sources record as containing 50 cubic feet of timber were, for the most part, probably around 60 or 70 years old; those containing 25 cubic feet were around 35 to 40 years old; although in both cases slightly older where Hoppus feet were being used.

Working on this basis, it is clear that most timber trees growing on farmland in the period before the mid-nineteenth century were felled before – and often long before – they were middle-aged. This can be illustrated by using data from one county, Essex. At Elmstead in 1724 two groups of trees were valued; one comprised 24 examples and had an average volume of only 9.3 cubic feet; the other, 121 trees in all, had an average of 12.4 cubic feet (ERO D/DR E5). Hedgerow oaks felled on Slough House Farm in Purleigh and Danbury in 1827 had an average volume of only 6.9 cubic feet and those taken down in 1834 an average of only 7.5 cubic feet, while oaks felled in 1835 averaged 14 cubic feet, and one small group only 3 cubic feet (ERO D/DOp E4). Of the 15 groups of ash trees surveyed on the Toppings Hall estate in Hatfield Peverel in 1791, a total of 691 trees in all, only a third included any examples with volumes greater than 50 cubic feet. The average volume for most parcels was less than 15 cubic feet: only one (comprising four trees from Mead Plain) had an average above 35 cubic feet (37.5), skewed by one abnormally large tree of 61 cubic feet (ERO D/DRa E23/5). Most striking of all are the figures from a survey of the

1,214 timber trees growing (mainly in hedges) on an estate in Waltham Abbey in 1791. Of the 762 oaks surveyed, only 2 per cent were estimated to contain more than 25 cubic feet, and none more than 40; all of the 255 ash trees were thought to contain less than 15 cubic feet; and, while some of the 197 elm were larger, one containing an estimated 40 cubic feet, most contained less than 15 (Ipswich RO HA 116/5/11/2). The vast majority of timber trees in the Essex countryside were thus being taken down well before they were 60 years of age.

Similar patterns are apparent in other counties. On the Chevallier estate at Aspall in Suffolk a 'valuation of the oak timber began to fell 22 April 1728' gives details of 57 trees, most apparently from hedgerows. The majority (53 per cent) had volumes of less than 20 cubic feet; a further 33 per cent contained less than 30 cubic feet; only one tree contained more than 40 cubic feet. The girths tell a similar story: nearly a half had quarter-girths of less than 10 inches, equivalent to actual girths of around a metre; only one tree had a quarter-girth greater than 15 inches. Another list, from 1736, describes four trees with volumes of less than 20 cubic feet, five in the 20–30 range and four above 30; the largest, however, contained only 34 cubic feet (private archive). In c.1830 the six 'best' trees on Thurston Hall Farm in Hawkedon in the same county had volumes ranging from 34 to only 60 cubic feet, suggesting that most of the timber on the property was smaller than this (Bury St Edmunds RO HA535/5/35). In Norfolk the average volume of trees felled on an estate at Sparham, Bintree and Billingford in 1796 was 28 cubic feet, while on the glebe lands of Merton in 1846 only eight of the 68 oaks contained more than 40 cubic feet of timber; even the largest, with a volume of 64 cubic feet, was not necessarily much older than 70 years (NRO WLS LXX/22/1–41, 481x7). The average volume of 762 oak trees growing on the Stow Bardolph estate in Stow and Wimbotsham in 1815 was 18.1 cubic feet, and that of oaks growing in Shouldham Thorpe, Shouldham, Marsham and Fincham a mere 10 cubic feet (NRO HARE 5500, 223 x 1). The 85 oaks growing at Hindolveston in 1704 had an average volume of around 25 cubic feet (NRO DCN 59/21/3), while on an estate in Hunworth, Stody and Brinningham in north Norfolk in 1739 the average volume of 2,983 oaks was just under 11 cubic feet (NRO NRS 16112, 32A6). In survey after survey, from all over the country, the same pattern is repeated. The average size of standing timber on a property at Harpenden in Hertfordshire in 1775 was 31 cubic feet (NRO 20757A); on a farm in Chiddingstone in Kent in 1766 80 per cent of the mature oaks were under 20 cubic feet and only 3 per cent were over 40, the ash trees had an average volume of only 6.2 cubic feet and the elm of 15.2 (KHL U908/E37); while on a property at Bircher in Herefordshire a valuation made in 1828 recorded 317 oak trees with an average volume of 15 cubic feet and 187 ash with an average volume of 8 cubic feet (HAR F76/111/50).

Perhaps the point has now been firmly made, but it is hard not to add the evidence from the Boughton estate in Northamptonshire, where the volumes of around 20,000 farmland trees were recorded in 1749. Only 1.8 per cent of the oaks growing in the hedges of the township of Weekley in 1749 contained more than 20 cubic feet of timber. On Warton manor the figure was about 5.5 per cent; on Newton manor, 5.7 per cent; in Stoke

Doyle, 5 per cent; in Little Oakley and Brigstock, 4.3 per cent; in Geddington, 5.8 per cent; in Kettering, 3.3 per cent; in Kingsthorpe, 2.7 per cent; in Hanging Houghton, 2.2 per cent; in Barnwell, 1.8 per cent; in Hemington, 0.9 per cent; and in Armston a mere 0.8 per cent (Boughton House archives). On the whole of the estate there were no ash trees or elm trees at all with volumes in excess of 20 cubic feet.

It should be noted that there was some variation in the size of the trees of different species, either recorded in surveys or measured when felled. In general, elms tended to be larger than oaks, which were in turn larger than ash. On a property at Little Walden in north-west Essex in 1826, for example, the 32 elm trees surveyed had an average volume of 7.6 cubic feet, but the 313 oaks averaged 6.5 and the 93 ash only 4.9 (ERO D/F 35/3/42). At Finchingfield, also in north-west Essex, in 1773 the 116 elm trees recorded had an average volume of 18 cubic feet (the largest estimated at 67); the 1,377 oak averaged 13 cubic feet (the largest 50); and the 90 ash averaged only 7 cubic feet (the largest containing 20) (ERO D/DPg T8). This pattern probably reflects the uses to which different kinds of timber were put. As explained in Chapter 3, elm was mainly used to provide boards and oak for major structural timber, while ash was employed for smaller stuff, such as scantling timber, fencing and gateposts.

We should also note, perhaps, that average figures tend to obscure the fact that recorded populations sometimes included a few trees of much larger dimensions. On a property at Clehonger in Herefordshire in 1820 the average volume of the 520 trees recorded was only 22 cubic feet, but a few of the hedgerow oaks contained over 90 cubic feet (HAR C38/49/3/1). Similarly, the timber surveyed on the Panshanger estate in Hertfordshire in 1719 had an average volume of 29 cubic feet, but the largest example, clearly a very large and old tree, reached 315 cubic feet (Austin 2008). A tree of unknown species felled at Easton Lodge estate in Suffolk in c.1800 contained 130 cubic feet, while 72 trees hedgerow trees measured prior to felling at Kirton in the same county in 1769 had an average volume of 31.4 but included one example containing as much as 108 cubic feet (Ipswich RO C/5/1/1/8/40). A group of 126 oaks and 195 ashes growing on farmland at Newburgh on the edge of the Howardian Hills in Yorkshire, surveyed in 1774, mainly had volumes of 20 to 30 cubic feet, but some oaks were allowed to grow larger, with a significant proportion above 50 cubic feet, resulting in a mean average of 46 cubic feet, while clusters of trees around 100–120 cubic feet were also represented (Nrth ZDV/V/I [MIC 1282/7159]). But, while everywhere present, timber trees older than 50 or 60 years were in a very small minority in the countryside. In 1794 Pringle could note of Westmoreland that it was 'general opinion in this and, I believe, in other counties that it is more profitable to fell wood at fifty or sixty years growth, than to let it stand for navy timber to 80 or 100' (Pringle 1794, 12); Stevenson in 1809 described how in Surrey 'few oaks are suffered to reach 60 years before they are felled' (Stevenson 1809, 441).

Of course, this does not necessarily mean that the English countryside in the seventeenth and eighteenth centuries was everywhere filled with young trees, for as we have seen in most districts the majority of farmland trees were pollards. We might reasonably assume that these were more likely to attain a ripe old age, for they continued to produce a regular crop of wood and thus maintained their usefulness and value for centuries. But while it is true that pollards were, in general, allowed to grow for longer than timber trees they, too, were often replaced with younger specimens before they reached any very great age, in this intensively managed countryside. Thomas Hale in 1756 was adamant that the farmer should 'fell or stubb them up at the proper time'. This was because 'Pollards usually, after some Lopping, grow hollow and decay … . The Produce of their Head is less, and of slower Growth.' He argued that they should be taken down before the trunk rotted badly and lost value. Felled in time, their trunks would produce a good crop of firewood and perhaps some wood suitable for 'mechanical purposes'. He advised the farmer to ensure a constant succession by regularly replacing old pollards with young trees destined to be managed in the same manner (Hale 1756, 141). While neglect clearly allowed a proportion of pollards to reach a venerable old age – the ancient 'veterans' of today – such trees were always exceptions, and it is striking that one eighteenth-century observer, railing against the antiquity of the pollards cluttering up the hedges of East Anglia, commented disparagingly that these were 'of every age, under perhaps two hundred years', clearly suggesting that there were few any older than this (Middleton 1798, 345). Trees of two or even three hundred years' growth would scarcely be classed an 'ancient' today. In this context, it is striking how records from the nineteenth century suggest that while big old 'veteran' trees were more common than they are today they were still comparatively unusual. James Grigor, writing in 1841 about Norfolk, thus singled out for remark trees such as the Great Oak of Thorpe near Norwich, 'One which we may class with the most noted in our county', although it had a circumference of only 16 feet 3 inches, less than five metres; hardly any of those he noted had girths of more than 7 metres (Grigor 1841, 53). Most veteran oak pollards surviving today, both in Norfolk and more generally, have girths of less than six metres and, while they may seem ancient to us, the majority were perhaps less than two centuries old when active pollarding ended in the late eighteenth century. At that point they were probably in imminent danger of being removed by landowners and replaced by new, vigorous specimens.

Conclusion

Before the middle decades of the nineteenth century many areas of England seem to have boasted vast numbers of farmland trees, especially the old-enclosed counties of the south-east and East Anglia and (although perhaps to a lesser extent) the west Midlands. Densities of 20 trees per hectare were common; of 30, 40 or more not unusual. Even in unenclosed, 'champion' districts, more trees existed in the landscape than we usually assume, crammed into village enclosures, scattered along roadsides and across meadows, even on occasions standing in the middle of open-field furlongs. As the open fields were enclosed through the seventeenth and eighteenth centuries the numbers of farmland trees in these areas increased further, although they seldom reached the levels attained in old-enclosed areas.

But it was in northern areas that, so far as the evidence goes, the numbers of farmland trees were often low, and became lower over time.

Farmland tree populations before the nineteenth century also differed from modern ones in other fundamental ways. In most districts there was a very high proportion of pollards; the vast majority of timber trees were less than 50 years of age, and even pollarded specimens were probably, for the most part, less than two centuries old. Why trees existed in such numbers, and why tree populations were managed in these highly intensive and unnatural ways, must now be explored.

CHAPTER THREE

The economics of farmland trees

The artificiality of tree populations

The sheer numbers of farmland trees to be found in so many parts of pre-industrial England are remarkable, not least because – as contemporaries were well aware – they unquestionably interfered with the practice of farming. Smith, for example, suggested that they were not good 'either for Corn or Grass' in the adjacent fields, and weakened the hedges in which they grew (Smith 1670, 30). Timothy Nourse went so far as to assert that open fields were generally held 'in greater esteem' than enclosures partly on the grounds that 'Corn never ripens so kindly, being under the Shade and Droppings of Trees; the Roots likewise of the Trees spreading to some distance from the Hedges, do rob the Earth of what should nourish the Grain' (Nourse 1699, 27). Hedgerow trees, he suggested, also provided a home for birds that would predate on grain crops. Such concerns are echoed in the comments of local farmers and land managers. In 1740 at Thorndon in Suffolk the agent bemoaned the 'pollard trees which this estate is very much encumbered with', noting that 'if a good deal more were cut down, it would be much better for the land' (Bury St Edmunds RO BT 1/1/2). At Badwell Ash in the same county in 1762 it was said that the land was 'capable of great improvement by destroying the timber and pollards that encumber the fields in many places' (Bury St Edmunds RO B E3/10/10.2/28). Such objections were particularly strong in grain-growing districts, but even in pasture areas some farmers could see the disadvantages of too many hedgerow trees. Billingsley reported that in Somerset the graziers preferred large enclosures and objected to the planting of trees on the grounds that they 'harbour flies, which teaze the cattle, and check their progress in fattening; trees also prevent a free circulation of air' (Billingsley 1794, 175). By the second half of the eighteenth century many agricultural writers were advocating that trees should be removed altogether from hedges, and planted in woods and plantations only.

It should also be emphasised that trees in hedges were there by choice. Most were deliberately planted, rather than arising spontaneously. This is clear from the many documentary references to tree-planting, especially but not exclusively when new hedges were being established. The survey made in 1758 of the Duke of Powis' properties in Northamptonshire, for example – comprising a number of parishes that had seen much recent enclosure – contains such phrases as 'Fine young Quick fences of *c.*8 yrs growth with great number of thriving elm and ash trees', and 'Well fenced with young Quicks of *c.*8 years growth in which are planted a great many fine thriving trees' (NHRO ZB 1837). Moreover, in many places the planting of trees in hedges, new or old, was a condition of a tenancy. On the Temple Newsam estate near Leeds, for example, leases making this stipulation are regularly found from as early as 1608, continuing until at least 1768 (Lds WYL100/NE/2; Lds WYL100/NE/110). A typical example from 1709 bound the tenant to plant six oak, ash or elm trees annually for the 21-year duration of the contract (Lds WYL100/NE/83a). A seventeenth-century lease for Lannock Manor in Weston, Hertfordshire, similarly instructed the lessor to plant 'ten young trees every year (oak, ash or elm), cherish and preserve them' (HALS DE/B513/E1). In 1700 Lord Fitzwilliam instructed his steward to inspect all the 'wood growing which his tenants have planted and do plant yearly' on his estates in Northamptonshire (Hainsworth and Walker 1990, 44). Such provisions were found all over the country. A lease for land in Great Canford in Dorset, drawn up in 1725, instructed the lessee to plant six trees of oak, ash or elm every year; another, drawn up ten years later for land in the same parish, ordered that five 'young oak, ash or elm trees' should be planted annually (DRO D-WIM/JO-313B; D-WIM/JO-419A). In Troutbeck in Cumberland, similarly, seventeenth-century leases instructed tenants to plant oak, ash and elm trees on their holdings (CRO (C),D/Lons. 5/2/11/283).

Indeed, even in the Middle Ages there are references to the planting of farmland trees. At Hindolveston in Norfolk in 1312 two men were paid for 'pulling ashes to plant at Hyndringham [Hindringham] and Gateli [Gateley]' and one for 'planting ashes in the manor' (Rackham 1986, 224). In Forncett in the same county in 1378 men were paid for 'pulling plants of thorn and ash to put on 1 ditch from the south of the manor to the churchyard' (Davenport 1906). Not surprisingly, seventeenth- and eighteenth-century writers on farming and forestry generally assume that trees were deliberately planted in hedges and on farmland, rather than arriving there adventitiously. Ellis always refers to elm as 'planted in a hedge'; Blagrave urged the raising 'upon each Lordship or Pasture, [of] Fuell and Fire-wood sufficient to maintain many Families, besides the Timber which may be raised in the Hedge-rows, if here and there in every Pearch be but planted an Ash, Oak, Elm' (Blagrave 1675, 114). At the end of the eighteenth century William Marshall, writing about Norfolk, described how 'Upon some estates it is the practice to put in, when a new hedge is planted, a holly at every rod, and an oak-plant at every two or three rods, among the whitethorn layer'; while Nathaniel Kent described how small proprietors, enclosing open-field land piecemeal in the same county, set within the new hedges 'an oak at every rod distance' (Marshall 1787, 113; Kent 1796, 72). Mortimer in 1707 thought that 'The best way of raising Trees in Hedges, is to plant them with the Quick': but he also gave advice on how to establish them 'wheare Hedges are planted already, and Trees are wanting' (Mortimer 1707, 309). Tuke, writing about Yorkshire at the end of the eighteenth century, worried about the 'neglect of planting trees in hedge-rows, and of proper management of those which are now growing there' (Tuke 1800, 191).

While ash will seed itself readily in hedges, elm will only sucker from the roots of trees already planted there, while oak – the main hedgerow tree across large parts of the country – will not readily establish itself in hedges (as opposed to woods) on its own. The evidence provided by the many surveys that have been made over the last 50 years of the number and varieties of shrubs growing in hedges, from all over the country, accordingly show that ash is frequently present, apparently self-seeded, in hedges established in the eighteenth and nineteenth centuries, but that oak is rare as a shrub, although frequently present as a *tree* – clearly indicating that most of those now growing in hedges have not developed from self-seeded specimens but were instead deliberately established there (J. Hall 1982; Addington 1978; Cameron and Pannett 1980). Even when oaks did seed themselves in hedges, intentionally planted specimens were generally preferred. As Rudge put it in 1813, oaks 'require to be raised from the acorn, or transplanted and protected; for although they will re-produce themselves … yet, growing slowly, the young shoots, or plants, seldom rise above the hedge-rows, so as to become trees, or if they do, are galled, knotty, and stunted' (Rudge 1807, 367). Where hedgerow trees were self-seeding, moreover, they had to be carefully protected from grazing stock and other hazards, including damage by tenants, who, as we have noted, had little interest in preserving timber that might interfere with their farming. Initial establishment, that is, might be the consequence of natural processes, but the plant's survival to maturity was a function of human agency. A lease for a farm in Barnet in Hertfordshire, drawn up in the early eighteenth century, typically instructed the tenant to 'do every Thing in his Power for the Encouragement, and growth of the young Timber Shoots, under the Penalty of Twenty Shillings for every Shoot or Sapling which shall be wilfully hinder'd from growing', with a penalty of 20 shillings prescribed if he was to 'stub up, prune, or injure any Sapling', and of five pounds for destroying or injuring any young timber tree (HALS DE/B 983 E1).

The incredible densities of farmland trees that we often encounter in maps and documents from the period before the mid-nineteenth century were thus the consequence of deliberate choices made by owners, tenants or both: choices made in the full knowledge that hedges densely filled with trees adversely affected the production of crops and livestock, which was the main business of the land. What is particularly striking is that the density of farmland trees and the mode of their management appear to have been broadly correlated: the higher the density of farmland trees, on the whole, the higher the proportion that were cropped as pollards. At one extreme, most northern counties such as Yorkshire – at least by the eighteenth century – appear to have had few pollards, and densities of trees that seldom, even on enclosed land, rose above *c.*4 per hectare. At the other, in old-enclosed parts of Hertfordshire and Norfolk over 70 per cent of trees were pollarded and there were generally more than 20 farmland trees per hectare, and often more than 30 (Figure 3.1). Midland counties such as Northamptonshire – where pollards generally comprised around 20–25 per cent of trees in enclosed parishes, and there were rarely more than 10 trees per hectare – fall somewhere between these extremes. In fact, differences in the overall density

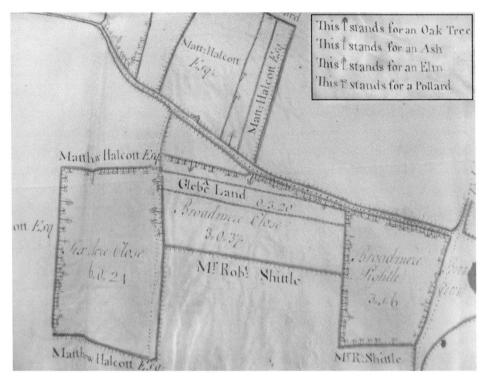

3.1 Another section of Henry Keymer's 1764 map of a farm at Beeston-next-Mileham in Norfolk. Note the high proportion of pollards.

of farmland trees across enclosed land are largely, although not entirely, accounted for by variations in the numbers of pollards; the density of *timber* trees across different regions displayed more consistency.

The uses of pollards

What mainly needs to be explained, therefore, is the large number of pollards present in many areas, and this in turn raises the question of what the material cut from pollards was actually used for. A number of historians and ecologists have highlighted the importance in the pre-industrial economy of 'leafy hay': that is, material cut from trees – especially ash, holly and elm – to provide winter fodder for livestock. Pollarding for fodder was certainly a widespread practice in other areas of Europe (Read 2008), but in England the majority of references come from the north of the country. Thomas Pennant thus described how, on a tour through Cumberland in 1772, he saw 'that the tops of all the ash trees were lopped; and was informed that it was done to feed the cattle in Autumn, when the grass was in decline; the cattle peeling off the bark for food' (Pennant 1776, 33). The right to lop young ash branches to provide fodder was cited as the reason for the lack of good timber on many Cumbrian manors by the sixteenth century, and Fleming has suggested that the practice was also well-established in parts of the Pennines, especially Swaledale, where it may have continued on a casual basis into the twentieth century (Fleming 1998). Many early references are probably to trees growing on commons, rather than in field boundaries, but certainly not all, and it is noteworthy that the court of the manor of Windermere in Westmoreland (now Cumbria) ruled in the seventeenth century that it was an offence to 'cutt downe or breake any other Men's Ash leaves', suggesting that the trees in question were growing on enclosed property (Rollinson 1974, 84). As we shall see, holly was also grown and pollarded for winter fodder in many northern districts, and on some scale, although mainly in special enclosures or on common land, rather than in among the fields (Spray 1981).

Yet while their role as providers of fodder is often emphasised, even in northern areas pollards were also used to supply other commodities. Furthermore, Gledhill's analysis of leases and accounts in Swaledale implies that leaf foddering had all but ceased in that district at least, by the seventeenth century (Gledhill 1994, 305–6). In the Midlands and the south, in the post-medieval period at least, there is little evidence for the practice. Winter feed was mainly if not exclusively supplied by hay cut from meadow land, until the widespread cultivation of turnips began in the later seventeenth century. Here, references to the use of trees for fodder are largely restricted to forests and parks: as late as 1790 ash trees were regularly being lopped as deer feed in Rockingham Forest, for example (*Journal of the House of Commons* 1792). It is true that Thomas Tusser, who was born in Essex and who farmed for much of his life on the Suffolk–Essex border, advocated feeding the cuttings from farm pollards – and the trimming from timber trees – to cattle during the winter months. However, his wording implies that this was as a by-product of cutting for fuel, the cattle presumably eating the twiggy growth and leaving the rest for firing (Tusser 1573, 4). Leaf fodder may have been more important in some southern regions in the Middle Ages and it remained significant in parts of the north well into the post-medieval period, but the immense quantities of poles cut from farmland pollards were mainly used in other ways.

Some writers on agriculture and forestry suggest that poles could be used for a range of practical purposes around the farm, for fencing, building repairs or making tool handles. Davis, writing about Wiltshire at the end of the eighteenth century, bemoaned the apparent reduction in the numbers of ash trees in the county, which were 'so very necessary for implements of husbandry to the corn farmer' (Davis 1794, 90), while Parkinson advocated the planting of ash pollards in hedges as late as 1808 on the grounds that they would eventually provide 'very useful farmer's wood for hurdles, rails, fences etc' (Parkinson 1808). Far more common, however, are references to the use of pollard poles as hedge stakes. Many hedges in England were managed by 'plashing' or 'laying' at intervals of ten or so years. After much of the hedge had been hacked back and the material removed, the major stems were cut almost through at a point close to their base and then bent over to a near-horizontal position. They were woven between a small number of upright stems that had remained uncut and a much larger number of vertical 'stabbers' or 'hedge-stakes', which were hammered into the ground. Continuous 'cables' of twisted hazel called 'edders' or 'ethers' ran along the top of the completed hedge to bind it all together. As the stems

3.2 Typical laid hedge, showing 'stabbers' or hedge stakes, and the 'ether' – the cable of twisted hazel running along the top of the hedge.

recovered and sprouted new growth, the hedge became a dense green wall, impenetrable to stock. This form of management, practised over much of England, required very large numbers of stakes made of straight but sturdy pieces of wood of the kind that pollards produced (Figure 3.2). Even where hedges were managed (as they often were) simply by coppicing, significant numbers of such stakes were required for the 'dead hedges' of thorn and twigs that were constructed to protect the regenerating stools. The use of pollard poles for such purposes is often referred to in early documents. For example, a lease for 21 years, drawn up in 1632 for a farm at Colchester St Peter in Essex, allowed the tenant to take 'hedge bote and stake bote' from the tops of the 'bowlings' (pollards) growing on the premises (ERO D/P 178/3/14). A survey of the Blickling estate in Norfolk, drawn up in 1756, described one farm in Wymondham on which there was 'Little or no Timber & scarcely Pollards enough for hedge stakes the Tenant is therefore obliged to buy Coals. It lies within two miles of Norwich' (NRO MC 3/252, 468X4).

As this quotation implies, however, pollards were also used to supply firewood and, indeed, most seventeenth- and eighteenth-century commentators discuss pollards simply in terms of fuel (Warde and Williamson 2014). Nourse, for example, described how hedgerow oaks, elms, ashes 'and the like' could be lopped for fuel (Nourse 1699, 28); Blome discussed 'those *Trees* not fit for *Timber*, but such that are designed for present use, for *Fewel &c.*', which could be 'lopped or shredded, at convenient seasons' (Blome 1686, 254). Hedgerow trees provided a supply of 'Fewel, so deficient in many *Champaine Countreys*' (Blome 1686, 250). Poles cut from pollards were clearly used in other ways, for hedging and the like, but fuel was evidently uppermost in the minds of most contemporaries. This in turn suggests that variations in the numbers of pollards in any area, and thus to a large extent in the overall number of farmland trees, were closely related to the character of local fuel supplies. Where there was a particular demand for firewood the kinds of problem caused by high densities of hedgerow trees, noted above, had to be accepted. But in this context it is important to remember that wood was only one of the fuels available in England in the seventeenth and eighteenth centuries.

Fuel economies

At the end of the sixteenth century William Harrison argued that a shortage of wood in the vicinity of London would soon drive the inhabitants of the city to burn a wide range of alternative fuels, employed in other districts, including 'fenny bote' (peat), broom, turf, heather, 'furze' (gorse), bracken, flags, straw, sedge, reeds and rushes (Edelen 1994, 281). Many of these materials were derived from heathland, especially broom, gorse and heather ('ling'), the latter cut in the form of turves dug to a depth of at least 2.5 cm, which thus included both the vegetation and, more importantly, a portion of matted root material. All were used as domestic firing, but they were also employed industrially, especially for firing brick kilns. When Blickling Hall in Norfolk was constructed in 1617–21 more than a million bricks were fired in kilns entirely fuelled with gorse and broom faggots brought from the nearby heaths, and well into the nineteenth century most of the kilns on the Bedfordshire brick fields were likewise fired using heathland vegetation (NRO MC3/4; Cox 1979, 27). It has even been suggested that the vitrified bricks so common in seventeenth-, eighteenth- and even some nineteenth-century buildings throughout England, and often employed in a decorative fashion, may have been the consequence of using such fuel: 'the high proportion of vitrified headers was probably a consequence of using firing materials, such as heather and gorse, which gave off fumes containing potash' (Cox 1979, 28–9). Where demand was high or alternative firing in short supply, heathland vegetation might be carefully nurtured. Indeed, in the south-west of England gorse was often grown in private enclosures. Even in 1801 it was said of Sithney in Cornwall that 'Here are, it is almost literally true, no trees; consequently a considerable part of every estate is under furze, which would frequently, with proper cultivation, produce whatever the cultivated lands now produce' (Turner 1982, I, 33–4).

Most heaths were common land, however, where the production of heather and gorse might be limited by the intensity of grazing. John Norden, writing in 1608, described the gorse in the West Country, which grew 'very high, and the stalke great, whereof the people make faggots'. He continued:

> And this kind of Furse groweth also upon the Sea coast of *Suffolke*: But that the people make not the use of them, as in *Devonshire* and *Cornwalle*, for they suffer their sheep and cattell to browse and crop them when they be young, and so they grow too scrubbed and lowe tufts, seldome to that perfection that they might be. (Norden 1608, 235)

Careful management, often involving the 'doling' of certain areas to commoners – that is, allocating portions of heath to individuals or families in the forms of strips – might therefore be necessary in order to ensure a fair division of the combustible vegetation. Alternatively, areas of common heath might be communally harvested in turn and then fenced off until the vegetation had recovered, as seems to have happened on Bromley Heath in Essex in the 1620s (ERO D/DU40/96). Where heaths were managed in these ways they could provide a rich source of fuel. In the early seventeenth century Thomas Blenerhasset could comment of Horsford Heath, near Norwich, that 'This heathe is to Norwich and the Countrye heare as Newcastle coales are to London' (Barrett-Lennard 1921, 120).

Harrison's 'fenny bote', or peat, was a far more important fuel source. In East Anglia the progressive drainage of the Fens during the seventeenth and eighteenth centuries greatly reduced the availability of peat, although smaller deposits continued to be exploited in areas such as the Norfolk Broads, and even in south Cambridgeshire peat continued to be dug on some scale well into the twentieth century (Day 1999, vi). In Lincolnshire, by contrast, and on the Isle of Axholme especially, it remained a significant fuel for much longer, Eden reporting in 1797 that it was 'the usual fuel consumed by labourers' (Eden 1797, 566). But it was in upland districts, where raised bogs supplied ample reserves of *sphagnum* peat and where extensive lowland mosses also often existed, that peat retained its importance as a fuel throughout the post-medieval period (Rotherham 2011). In the later eighteenth century peat was still being removed on such a scale from the upland commons in parts of the north-west that it was causing serious damage to the grazing. At Bolton in Westmoreland in the early nineteenth century, for example, it was argued that if the common was not enclosed it would soon be completely ruined by peat digging (Whyte 2003, 33).

Like heathland vegetation, peat was employed industrially as well as domestically, usually in the form of peat charcoal. It was used to fire lime kilns and, alongside wood charcoal, for smelting tin and other non-ferrous metals, only being replaced by coal in the first half of the eighteenth century following the development of the reverberatory furnace (Palmer and Neaverson 1994, 128). The extent to which peat remained a major source of fuel in upland districts is clear from the way that parliamentary enclosure awards often allocated, to those receiving allotments, a 'moss dale' where they could cut peat (Whyte 2003, 76). These continued to be made into the middle decades of the nineteenth century, as at Troutbeck in Westmoreland in 1840, and were exploited by farmers as much as by smallholders and cottagers. In more remote areas of the Lake District peat was still being widely consumed even in the 1930s (Rollinson 1974, 102–3). Upland moors also supplied other forms of fuel: heather was the dominant vegetation on many of those in the east of the country, or at lower altitudes, where the deposits of peat were usually less significant.

The other key fuel source in post-medieval England was coal. This was in widespread use as domestic firing in many parts of England before the start of the seventeenth century, but mainly on or near the main coalfields, which were located in Shropshire, Staffordshire, south Yorkshire, Derbyshire and, above all, in the north-east of England. Away from coalfields it was burned on any scale only in places to which it could be brought by boat, the transport of such a dense commodity over long distances being expensive and difficult. William Harrison, writing in the 1570s, noted how the use of coal was at that time just beginning to spread 'from the forge into the kitchen and hall, as may appear already in most cities and towns *that lie about the coast*, where they have little other fuel except it be turf or hassock' [our italics] (Edelen 1994, 281). This importance of water transport was still being emphasised by Pehr Kalm in 1748, when he observed that coal could be found in London and was widely burned in villages within a 14-mile radius of the city, but 'in places to which they have not any flowing water to carry boats loaded with coals' the population continued to burn wood – mainly from 'trees they had cut down in repairing hedges' – or 'fuel of some other kind, as bracken, furze etc'. Only gradually, as road transport was improved and canals were constructed in the course of the eighteenth century, did its use extend more widely, although even then often only among the wealthier elements of society. A government survey of 1790 described how coal was the most important fuel in Devon in the 'Houses of Creditable People, but the poor burn no Coals, and very little Wood, on Account of the Expense ... most of their Fuel is Turf or Peat' (*Journal of the House of Commons* 1792, 328). It was only in the mid-nineteenth century, with the spread of railways, that coal became the normal industrial and domestic fuel throughout England.

The character of fuel supply, then, was almost certainly a crucial factor in determining the numbers of pollards that people were prepared to tolerate on farmland, and thus to a large extent the overall density of farmland trees. Where coal or peat was abundant – and perhaps where heaths were extensive and lightly grazed – farmers and landowners would have been less keen to fill the hedges with trees. But the situation was complicated by the presence of large urban centres and the demand for fuel that they provided: places such as London, or even Bristol or Norwich, critically shaped the character of land management in their hinterlands. The proximity of London, for example – coupled with the poor character of the soils formed in London Clay and pebble gravels – almost certainly explains the particularly high densities of pollards found in the south of Essex and Hertfordshire and in the north of Surrey. Complex patterns of supply and demand, variable over time and space, thus underlie the variations in the character of farmland tree populations revealed by maps and documents.

In the north of England the density of farmland trees and the proportion managed as pollards were both comparatively low by the eighteenth century simply because many communities had easy access to local reserves of peat, and peat was also transported significant distances, especially to major towns and cities. In the sixteenth and seventeenth centuries a large proportion of the inhabitants of York burned peat brought from Inclesmoor, nearly 30 kilometres away (Hatcher 1993, 124). Peat was similarly extracted on a commercial scale in the mosses of south-west Lancashire well into the eighteenth century and supplied to Ormskirk and Liverpool (Langton 1979, 56–7). In many districts, moreover, coal was widely used as a domestic fuel by the start of the seventeenth century. Indeed, as early as the 1530s John Leland was able to describe how, although wood was plentiful across much of Yorkshire, many people were burning coal (Toulmin-Smith 1907). By 1790 it had become almost the only fuel consumed in Durham, Yorkshire, Nottinghamshire, Staffordshire, Lancashire and Cheshire (*Journal of the House of Commons* 1792, 328–9). In the West Riding of Yorkshire 'The Use of Coal … has been universal, as far back as can be remembered'; in Staffordshire 'Coals are, and have been universally used in this county'; in Durham there was 'so much coal in all parts of the County … that no Wood is used as Fuel'; and in Nottinghamshire 'Coal of various kinds have always been so easily obtained … that the use of Wood for Fuel has never been considerable.'

We noted in the previous chapter how in most Midland 'champion' districts, while trees proliferated as more and more land became enclosed in the course of the seventeenth and eighteenth centuries, they did not reach the kinds of densities seen in old-enclosed counties such as Hertfordshire or Norfolk. Nor did the proportion of such trees that were managed as pollards reach similar levels. To an extent this was probably because, in some areas, extensive forests that provided a significant supply of fuelwood survived well into the post-medieval period, including Rockingham, Salcey, Bernwood, Whittlewood and Charnwood. But it was mainly because many Midland districts were located a relatively short distance from active coalfields. Much of Northamptonshire, for example, lay less than 25 kilometres from the Warwickshire coalfield, and less than 50 from the Staffordshire mines, both producing coal on a significant scale by the seventeenth century, while coal could also be imported from north-east England along the rivers Nene and Ouse. Morton in 1712 emphasised how 'the remotest part of the county is not twenty miles from the seats of coal, either of the inland or of the Newcastle sort, which is brought so nigh as Stamford, Bedford and Huntingdon by water' (Morton 1712, 16). Early leases for land in the county sometimes included a stipulation that tenants would, in addition to any cash paid for the use of a particular property, agree to transport coal for the lessor. One example, relating to land at Crick in 1582, described how the tenant was bound to cart one load of coals each year to the proprietor's house in Harlestone; another, from 1630 and for land in Hellidon, obliged the tenant to carry every other year 'from any place William Bradgate appoints not above 17 miles distant from Helidon to any place in Helidon or any place not more than 2 miles from Helidon one load of coals' (NHRO A/089; NHRO HOLT 655). Much of 'champion' Leicestershire similarly lay only 30 kilometres or so from the Derbyshire coalfields, while within the county itself mines at Cole Orton and Lound were operating by the seventeenth century and in the eighteenth century new mines were established by the Earl of Moira at Ashby Wolds.

In the period after *c*.1700, as more and more land was being enclosed and hedgerows more widely planted, access to coal was steadily improved across large areas of the Midlands by the development of transport infrastructure. In the case of Northamptonshire, parliamentary acts were passed in 1714 and 1724 to make the Nene navigable from Peterborough to Northampton: navigation was improved as far as Oundle in 1730 and, in 1761, as far as Northampton, all of which made sea coal more easily available. But road transport was also made better, with the turnpiking of major routes such as that from Warwick to Northampton (via Southam and Daventry) in 1765. In 1750 the inhabitants of Northampton were paying 30d per hundredweight for coal that cost only 4d at the pithead in Warwickshire; by the 1760s the price had fallen to 14d a bushel, around 21d per hundredweight (Hatley 1980).

In East Anglia and the Home Counties, by contrast, demand for fuel was particularly high because of the density of the population, but alternatives to firewood were generally in short supply. Although coal from the north-eastern mines was reaching coastal towns by medieval times, and much use was made of the resources provided by local heaths and fens, the market for firewood remained buoyant, leading to the high numbers of farmland pollards that we have described. Only where woods, and especially common wood-pastures, or tracts of heath and fen, were especially extensive were pollard densities noticeably lower – the Chilterns and their dipslope being an example of the former, north-east Norfolk of the latter. Even in these circumstances hedgerow pollards might be abundant, as, for example, in areas close to fuel-hungry London. Parts of south Hertfordshire and south Essex thus had – as we shall see – large areas of common wood pasture, even in the eighteenth century, but also a high density of pollards on farmland. It is striking in this context how some early writers in the Midlands and south of England seem to have associated pollarding closely with the proximity of the capital. Pearce, writing about Berkshire, lamented the 'shameful abuse of shredding and lopping of trees in the hedge-rows' that he observed in the east of the county, 'as is practised in the vicinity of London' (Pearce 1794, 57).

It might reasonably be objected that fuel economics cannot explain the substantial numbers of pollards that were to be found in the old-enclosed parts of the west Midlands, especially in Shropshire, where coal was already being extracted by the end of the Middle

Ages – although, as we have seen, there are indications that trees managed in this way were fewer than in the old-enclosed districts of the south-east or East Anglia. The output of the principal mines here was largely absorbed by local industries, especially after Abraham Darby began to use coke to smelt iron in 1709. Plymley noted in 1813 that coal was used as a domestic fuel in Shropshire, but described how peat and wood were also extensively burnt, while in Herefordshire, close to the drift-mines of the Forest of Dean, 'coal is in general use as fuel by as many of the inhabitants as can afford the purchase of it', although the high costs of transport away from the river Wye probably ensured that this was a minority (Duncumb 1805, 141). In short, the numbers of farmland trees and pollards within any area were, in broad terms, in inverse proportion to the availability of other sources of fuel, once the density of population and the demands of major cities are taken into account.

While the economics of fuel supply clearly played a fundamental role in generating variations in the numbers of pollards found in any district, it is likely that other factors were also important. Some of the material cut from oak pollards, for example, was employed as minor elements in timber-framed buildings, and it is noticeable that many of the areas in which in which pollards were particularly abundant – the south-east, southern East Anglia, Shropshire and Herefordshire – possessed no good building stone and thus had a strong vernacular tradition of timber-framed building. In such areas timber continued to be employed in the construction of outbuildings and the like well into the nineteenth century. In contrast, in areas where stone was readily available – that is, across most of northern England and in large parts of the Midlands – it became the main building material in the course of the seventeenth century and in consequence pollards were fewer and trees in general less numerous. Pollards, as we have seen, also provided the vast number of stakes required to maintain hedges and were, likewise, less necessary in many northern districts, where fields were bounded by drystone walls.

In addition to such essentially environmental and economic determinants, social factors – the character of estates and tenures – may also have played a part in shaping the numbers of pollards present in an area. The bolling or trunk of a hedgerow pollard was usually, as we have seen, owned by the landlord, but the tenant had the right to take the crop of poles. Where trees rose spontaneously from hedges – or had even been planted there – tenants had little interest in preserving them as *timber*, which would be the property of the landlord, and to them no more than a hindrance in terms of shade and other damage caused to crops. In spite of the injunctions contained in many leases, tenants were thus sorely tempted to cut or plash young trees with the rest of the hedge – or to convert them into pollards. 'It is in the farmer's interest, to make every tree a pollard' (Pearce 1794, 57). This would be more likely to happen where property lay in small or discontinuous and poorly administered estates with absentee owners. In this context we should note that some of the highest concentrations of pollards occurred on the claylands of East Anglia, Essex and Hertfordshire, where these conditions often pertained. John Middleton, while bemoaning the fact that 'many of the hedgerows' in Middlesex were 'disfigured by pollard trees', believed that the situation was worse in other counties. 'I never saw hedge-rows in any district so barbarously used by the tenants, or that reflect so much want of attention on the part of landlords, as those in Norfolk and Suffolk' (Middleton 1798, 275). In Essex, according to one commentator in 1790, 'the Landlord is of course desirous to encourage the Growth of Oak Timber in Hedge Rows: but the Farmer is so much used to make Pollards of very Thing, that he generally contrives to head young Trees, before they come to any Size' (*Journal of the House of Commons* 1792, 318).

We might note, in passing, that in certain contexts trees might have been pollarded for reasons other than to provide a supply of poles. Pollarding was also a way of controlling the height of a tree. In wetland areas pollarded willows were often used to stabilise river banks and, in particular – at least by the end of the eighteenth century – roads running across wetlands: the trees were planted to either side and their roots 'held the shoulder', providing a firm base for the road surface on spongy ground. In such cases the trees needed to be cropped regularly to ensure that they were not brought down by the wind, and the road in consequence ruined or the river banks damaged. Even where pollards were employed in this manner, of course, the material cut from them would have been used as fuel, or in other ways.

Hedges as collieries

In old-enclosed districts lying at a distance from coalfields not only the pollards growing in hedges but the hedges themselves were regarded as an important source of fuel. Thomas Tusser contrasted the 'champion' districts of England with the early enclosed 'woodland' areas, in which 'in every hedge' there was 'plenty of fuel and fruit' (Tusser 1573, 94). William Marshall noted in the 1780s how, in north-east Norfolk, the 'old hedges, in general, abound with oak, ash and maple stubs, off which the wood is cut every time the hedge is felled; also with pollards, whose heads are another source of firewood'. The entire supply of wood in the district, he added, 'may be said, with little latitude, to be from hedge-rows' (Marshall 1787, I, 96). In the Vale of Evesham, according to Turner in 1794, wood was 'scarce, the hedges and lop of trees being the chief supply' (Turner 1794, 45). Advocates of early enclosures in open-field areas often cited the improvements in fuel supply which would result from the planting of hedges (Reed 1981). Walter Blith claimed that if all land was enclosed 'No man almost in the Nation would be … at want of Firing' (Blith 1649, 113).

Seventeenth-century leases sometimes mention the tenant's right to take the 'offal' – the material cut from the hedge when it was coppiced or laid – for 'firebote' or domestic firing, but usually it was simply assumed that this would happen. For larger landowners and the larger tenant farmer, the firing cut from hedges seems to have been regarded largely as a means of covering the costs of maintaining them, although they too would have used a proportion of it, if only for their ovens. Randall Burroughs, an educated Norfolk farmer who kept a detailed diary of his agricultural work in the 1790s, regularly

3.3 An outgrown hedge newly disciplined by coppicing. Note the large amounts of fuel logs and brushwood produced. The upright posts are for the erection of a barbed wire fence to protect the regenerating stools from grazing livestock. Traditionally, a 'dead hedge' of staked thorny material would have been employed for this purpose.

paid his labourers for maintaining the farm hedges, in part, with the firewood cut from them (Wade Martins and Williamson 1995). When the Alscot estate in Warwickshire was put on the market in 1747 the sales particulars boasted: 'Note also that the hedges in the Liberty of Alscot will supply Alscot house with sufficient Firewood and pay the Charge of Cutting' (private archive).

It may have been demand for firewood that encouraged farmers in some districts to maintain their hedges by coppicing rather than laying them, for this produced a greater volume of wood (Figure 3.3). John Howlett, writing in 1807, described how hedges in Essex were 'usually cut down at the end of nine, ten or twelve years, and the pollard trees lopped'; they were cut to 'within an inch or two of the old stubb' (Young 1807, 180). The regrowth needed to be protected from livestock, usually by placing thorny material taken from the hedge over the newly cut stems. In many districts more complicated 'dead hedges' were constructed for this purpose, using hedge-stakes and even ethers, but in Essex at least this practice was on the wane by the start of the nineteenth century because, according to one commentator, they were 'sure to be torn up, destroyed, and burnt, stake, eathers, bushes and all, by the destitute poor, who from deficiency of wages, are utterly unable to purchase fuel, and compelled to steal it, or perish with cold' (Young 1807, 180). Whatever the precise form hedge maintenance took, it was usually carried out at the same time as the pollards growing within the hedge were lopped and the ditch running beside it – so vital for drainage on heavy land – was cleared of silt and vegetation. One lease, drawn up in 1693 for a farm in Aldenham in south Hertfordshire, stipulated that the tenant 'shall not lopp or cutt or cause to be lopped or cut any of the pollards growing upon the premises but when the hedges shall be new made and ditches scoured where the sayd pollards do grow' (HALS DE/Am/E3). Arthur Young recorded that hedges in Hertfordshire were generally plashed every 12 years, the rotation normally suggested by seventeenth- and eighteenth-century writers such as Blome. Howlett, writing in 1807, similarly described how hedges in Essex were 'usually cut down at the end of nine, ten or twelve years, and the pollard trees lopped' (Young 1807), but rotations as short as seven years are occasionally mentioned in leases and other documents (HALS DE/B983/E1).

Even hedges managed by plashing could produce a great deal of wood. Dormor has estimated (2002, 279–81) that the fuel requirements of eighteenth-century households in Wensleydale and Nidderdale could be met by the wood cut from 14–25 metres of hedge, while Thomas Hale in 1756 thought that 60 acres of enclosed land might produce a thousand faggots each year (Hale 1756, 119). When Randall Burroughs, the eighteenth-century Norfolk farmer–diarist, paid his workers for hedge maintenance he valued the cut material at 4d per rod (approximately five metres), a figure that excluded the wood cut from the pollards in the hedge, which he kept for his own use (Wade Martins and Williamson 1995). As he was paying, at the same time, 30 shillings for a cauldron (c.1.5 tons or 1,420 kg) of coal, we might estimate, very roughly, that the wood cut from 300 metres of hedge was worth roughly the same as a ton of coal. Obviously, the amount of usable wood produced depended in part on the kinds of shrub growing within a hedge, and it is noteworthy that Arthur Young described in the early nineteenth century how in Hertfordshire the acute need for firewood had 'induced the farmers to fill the old hedges everywhere with oak, ash, sallow and with all sorts of plants more generally calculated for fuel than fences' (Young 1804, 49).

It is perhaps mainly for this reason that hedges planted in the eighteenth or nineteenth centuries are today largely composed of hawthorn, with few other shrubs, while older examples tend to contain a large number of woody species. It is often assumed that this is entirely due to the gradual colonisation with new species of hedges originally planted with only one (Pollard et al. 1974), but while hedges certainly do acquire new species with the passing years it is noteworthy that eighteenth-century writers were already noting the difference between old mixed hedges and young ones of hawthorn. John Middleton, in his volume on Middlesex, contrasted the 'new quick hedges' and the older examples, which he described as 'consisting mostly of hawthorn, elm, and maple, with some black thorns, crabs, bryers and damsons' (Middleton 1798, 132–3). Boys, writing about Kent, noted the difference between 'old hedges, such as Nature has formed', and the newer 'quickset hedges raised from the berries of the white thorn' (Boys 1805, 62), while in Cheshire the contrast was between the new enclosures, of 'white, or haw-thorn', and the 'ancient fences', consisting of 'hasle, alder, white or black-thorn, witch-elm, holly, dogwood, birch &c &c' (Holland 1808, 122). And while some early writers, such as John Worlidge or Thomas Hale, advocated planting with thorn alone, in part because mixed planting made for a weak and gappy hedge (Worlidge 1660, 101; Hale 1756, 113), most describe and prescribe mixed planting (Johnson 1978, 197–9; Fitzherbert 1533, 53; Norden 1608, 201). Smith, writing in the late seventeenth century, thus believed that 'it be and hath been a general custom in *England*, to have several sorts of wood growing in hedge-rows' (Smith 1670, 27); Nourse similarly described how it was normal practice for farmers to sow, with the thorn, 'Acorns, Ash-keys, Crab-quicks and the like', not as timber, but to form the body of the hedge (Nourse 1699, 62), while Blagrave advocated the planting of hedges with species including elm, ash and oak. A number of studies of the character of hedges surviving in the modern landscape have similarly concluded that mixed-species planting was common, although not universal, before the eighteenth century (Willmott 1980; J. Hall 1982, 105; Barnes and Williamson 2006, 73–96).

The facts that hedges were regarded as a source of fuelwood and often included large quantities of ash, elm and, more rarely, oak; and that the pollards were lopped at the same time as the hedge in which they grew was plashed or coppiced; raise the question of why hedgerow trees were pollarded at all. It is usually suggested that coppicing was practised as a way of obtaining wood in contexts in which grazing livestock could be excluded and that pollarding was practised where they could not. But most pollards actually grew in what were, in effect, long and narrow linear coppices, from which browsing stock evidently could be excluded effectively enough to ensure the rapid recovery of the constituent shrubs. Why, that is, raise some plants above the level of the hedge, and cut them there, rather than simply coppicing them all together? The answer presumably lies in the kinds of material produced in 10 or 12 years by a regenerating hedge and by a pollard. Oak grows slowly and while it will maintain itself as a hedge shrub it will not flourish there, as we have seen (above, p. 55). The poles on an oak pollard, above the hedge, would put on much better growth. Even ash and elm would do better in such circumstances, raised above the dense shade of the hedge. Hedges would provide large amounts of bush faggots, suitable for ovens, kilns and the fires of the poor. Pollards provided, in addition, better quality firing, in the form of more substantial faggot poles, as well as material that could be used in a range of other ways: as, for example, stout hedging stakes.

Shaws and linear woods

Hedges, irrespective of the pollards growing within them, thus made a major contribution to the fuel requirements of old-enclosed districts, and in some areas they were evidently allowed to grow much wider than was required merely to provide a stock-proof barrier. Sir John Parnell asserted in 1769 that he knew:

No part of England more beautiful in its stile than Hertfordshire. Thru'out the oak and Elm hedgerows appear rather the work of Nature than Plantations generally Extending 30 or 40 feet Broad growing Irregularly in these stripes and giving the fields the air of being Reclaim'd from a general tract of woodland (LSE Coll Misc 38/3 f.8).

Some hedges of this kind were wide enough to be described in terms of the area that they covered. In 1556 there were 22 acres of 'hedgerows' on St Lawrence farm in Cheshunt in south Hertfordshire, ranging in area from one to four acres (PRO/TNA 315/391, fol. 144). At nearby Broxbourne in the mid-eighteenth century 'Whitestubbs Hedge Row', 'Pimbridge Hedge Row' and 'Palmer's Hedgerow' were all incorporated into the same 12-year coppicing rotation that applied to the woods and wood-pastures on the Broxbournebury estate, while, a short distance away, a map of Earl Cowper's estate at Hertingfordbury in 1704 depicts 'hedges' with green lines and 'hedge rows' as broader green strips containing trees (HALS DE/Bb/E26; HALS DE/P/P4). Similar features are recorded in the south of the neighbouring county of Essex. A survey of Cressing Temple, made in 1656, describes 16 'springs', or small coppices, bearing the same names as the adjacent fields (ERO D/DAc 96 and 101). The largest covered two acres, while most extended over a little less than one. All were managed as coppices, the survey describing the length of time since the underwood in each was last felled. A 1566 survey of the Belhus estate, Avery, in south Essex, described 176 acres of woodland, 144 of which comprised

3.4 Aerial view of the landscape of the Sussex Weald, showing the narrow strips of woodland or 'shaws' that form the boundaries of many of the fields.

'true' woods but a further 28 described as 'hedgerowes' (ERO D/DL M18), while at Newhall in Boreham in the same district it was reported in 1565 that a hedgerow covering half an acre had been 'fallen', or coppiced, six years previously but the regrowth had been damaged by grazing livestock (Rackham 1986, 189). Numerous examples of these features appear on early maps of the area, such as those of West Horndon Hall, made in 1598, or of Walthambury, made in 1643 (Hunter 1999, 137–41).

Similar wide hedges are recorded elsewhere. In 1810 Phillips referred to the 'broad irregular hedge-rows so frequently occurring in many parts of the county' of Hampshire (Phillips 1810, 293). They were common in the Chiltern Hills: in 1543–4 Henry VIII's woodward in Hertfordshire recorded the sale of eight acres of underwood 'from divers hedges in the parish of Flampsted called Hedgerowys' worth £10 13s 4d (PRO/TNA E315/457); an early eighteenth-century plan of an estate in nearby Great Gaddesden shows broad hedgerows bordered by trees extending around three fields, with narrower boundaries around the others; while a map of Herons Manor Farm in Wheathampstead, drawn up in 1768, similarly distinguishes these different kinds of field boundary (HALS 15594; HALSA DE/V/P2). But it was in Kent, Surrey and Sussex that narrow bands of coppiced woodland, here referred to as 'shaws', were particularly prominent features of the landscape, as to some extent they remain today, often outnumbering hedges of more normal form (Figure 3.4). In the Wealden parts of Surrey, according to Stevenson in 1809, 'almost every enclosure is surrounded by a broad belt of coppiced wood' (Stevenson 1809, 88). In Sussex it was reported that there were 'broad belts of underwood, and trees, two, three and four rods wide, around every petty enclosure' (Young 1808, 181). In the interior of the Weald most of the coppice had specialised uses, as we shall see, but towards the north much went as firewood to London or was converted to charcoal also destined for the capital.

The management of hedgerow timber

Apart from the high numbers of trees managed as pollards, the other key characteristic of the early modern countryside that requires explanation is the low age at which the majority of timber trees were felled, and thus the relative youth of most farmland timber. This was the consequence of a number of factors. The most important was that, prior to the development of large commercial sawmills in the middle of the nineteenth century, it was easier to select timber that was already roughly the correct size for a particular job than it was to allow trees on a property to grow to a large size and then saw them up to the dimensions required. Such an approach made particular sense given that, by around 70 or 80 years, the growth rate of oak, in particular, slows noticeably anyway. In addition, very large trees adversely affected the crops growing in adjacent fields, and in this context it is noteworthy that, where the evidence is sufficiently detailed, we can sometimes see that the largest specimens were to be found growing in meadows, pastures and parks, rather than in the hedges around arable fields.

The fact that bark is more easily peeled from younger oak trees, and is of a better quality for tanning, may also have been an important consideration. The demand for bark was a major influence on the management of trees from at least the late Middle Ages and by the late seventeenth century the tanning industry was being organised on a large scale. Consumption rose dramatically during the eighteenth century, and especially after *c.*1750, as the population grew rapidly and living standards improved. In the 1720s around 50,000 tons of bark were consumed by the industry each year in England, but by the early nineteenth century the figure was above 90,000 tons in most years (Clarkson 1974). The consumption of home-grown bark probably then stabilised around 1850, owing to imports, before declining dramatically as alternative methods of tanning, using in particular chromium sulphate and other chromium salts, began to be employed. The best bark for tanning came from oak coppice poles around 20 years old, or young oak timber of a similar age; the worst came from old trees, especially old pollards, which were also difficult to peel. In 1726 Daniel Eaton wrote that he had sold trees on one of the Northamptonshire properties of the Earl of Cardogan with an allowance 'for the bark at 6s 6d in the pound for the saplins, without any allowance being made for the doderells … for we are assured that a great number of them will not peel' (Wake and Webster 1971, 100–103). Bark was a significant source of income on Clement Chevallier's Aspall estate in Suffolk in the 1720s and 1730s (private archive). That stripped from 28 trees was sold in 1735 for £5 10s, equivalent to something like £800 today, while in the same year the bark from just four trees fetched 30s. In 1747 he sold the bark from 99 trees for the immense sum of £20 6s 8d. Bark was thus a valuable commodity, and that stored at a tannery described in the inventory of widow Judith Day of Hertford, taken after her death in 1688, represented 16.5 per cent of the total value of the business and was worth almost as much as all her household goods combined (Adams 1997, 100–103).

Conclusion

The most important observation we might make about farmland tree populations in the seventeenth and eighteenth centuries is that there was nothing very natural about them. Where trees grew and in what numbers, and how they were managed, were dictated by human choices and by practical, economic considerations. In particular, variations in the density of farmland trees and in the proportion managed as pollards were largely a function of the character of fuel supply. Firewood was one of several fuels available in seventeenth- and eighteenth-century England and where alternatives were readily available the incentive to pollard was less, especially if coal could be obtained at a reasonable cost. Pollards also supplied other useful materials, however, and where these were otherwise in short supply or in high demand – as with the hedge-stakes that were required in vast numbers in many old-enclosed districts – additional incentives for filling hedges with cropped trees existed. Similar practical considerations shaped other aspects of tree populations, especially the early felling of timber and thus the young average

age of the timber trees growing in fields and hedges. In innumerable ways, trees were embedded in local economies and in local social and tenurial conditions. And, while the particular character of these conditions may have been different in earlier periods, it had been very many centuries since trees had formed part of a truly 'natural' world, unshaped by human hand.

 CHAPTER FOUR

Woods and wood-pastures

The paucity of woodland

As the previous chapters have made clear, although much attention has been bestowed on enclosed, coppiced woodland by historical ecologists and others, there is no doubt that in most districts only a minority of trees grew in woods. Most grew on farmland, and especially in hedgerows. Indeed, by the time that reasonably accurate large-scale county maps were being published, in the middle decades of the eighteenth century, coppiced woodland was often a rare resource even in old-enclosed districts. Hertfordshire was, and is, a well-wooded county, but Dury and Andrews' county map of 1766 suggests that enclosed woodland accounted for no more than 10,096 hectares, a little over 6 per cent of its land area: although the map is not particularly accurate in this respect, this figure is probably broadly correct (MacNair *et al.* 2015, 106–7). The map of Norfolk published by William Faden in 1797 is more accurate, and shows that here enclosed woodland accounted for a mere 2.4 per cent of the land area (Faden 1797; MacNair and Williamson 2010, 119–25). It is clear that some woods had been lost from the landscape over previous centuries in both of these counties, as elsewhere. In the parishes of Great and Little Fransham in mid-Norfolk, for example, surveys made in 1603 and 1604 describe a number of small land parcels, generally covering less than 2 hectares, as 'late converted from wood to pasture', while a lease of 1637 describes the 2.5 hectares of Annyells Grove as 'being of late woodeground and stubbed up' (NRO MS 13167 40A7; NRO MS 13279 40C2; NRO MS 13159 40A6; NRO MS 13326 40C2). Archaeological surveys likewise suggest that many ancient woods had been truncated – and thus lost parts of their original boundary banks – by the time they were first mapped in detail in the seventeenth, eighteenth or early nineteenth century (Barnes and Williamson 2015, 76–80). On the other hand, existing coppiced woods were often extended in the course of the post-medieval period, again to judge from the field evidence, and some new ones established (above, pp. 22–4). Where the area of woodland shown on seventeenth-century maps can be compared with that shown on maps from the later eighteenth century, overall reductions are thus usually modest. It is very unlikely that, in most old-enclosed districts, more than 8 per cent of the land area had been occupied by coppiced woodland at the start of the seventeenth century. The only exceptions to this are the south-eastern counties of Hampshire, Surrey and Sussex, where the combined area of the numerous 'shaws', as well as larger woods, amounted – even in the nineteenth century – to more than 10 per cent of the surface area.

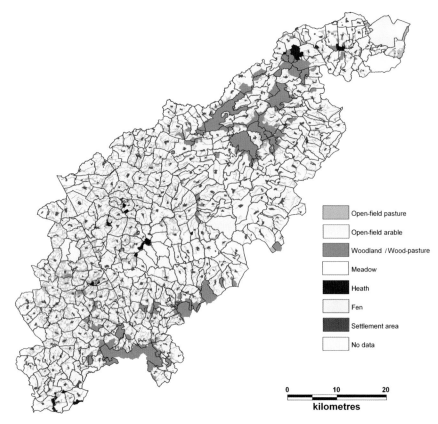

Open-field pasture

Open-field arable

Woodland / Wood-pasture

Meadow

Heath

Fen

Settlement area

No data

0   10   20
kilometres

4.1 Northamptonshire: a reconstruction of land use in c.1300, based on archaeological and documentary evidence. Most of the landscape is occupied by open-field arable, pasture and meadow. Woodland – including both enclosed coppice and wood-pasture – is strongly clustered in the three royal forests of Rockingham (to the north-east), Whittlewood and Salcey (south).

In marked contrast to the situation in these old-enclosed districts, many 'champion' areas had been largely denuded of woodland by the high Middle Ages and, in spite of some limited replanting, remained so into the eighteenth century. Leicestershire was said in 1794 to have 'very little timber … except in hedgerows' (Monk 1794, 56), while Batchellor estimated that there were only 7,000 acres of woodland in Bedfordshire, around 2 per cent of the county's land area. Even at the end of the nineteenth century, after a period in which landowners had planted numerous belts, plantations and coverts, the area covered by woodland in these districts remained small. In 1895 only 2.6 per cent of the land area of Leicestershire, 2.5 per cent of Lincolnshire and a mere 2 per cent of Huntingdonshire were occupied by woods and plantations. Most woods in the champion Midlands were concentrated within the royal forests, where they lay interspersed with areas of wood-pasture. In Northamptonshire, for example, enclosed woodland made up around 5 per cent of the land area in the middle of the eighteenth century, and over 90 per cent of this was located within the three royal forests (Figure 4.1). In the north of England there was even less woodland, concentrated – as noted in Chapter 1 – in particular locations. In Yorkshire, for example, it was mainly to be found on the margins of the main upland masses – the North York Moors and the Pennines. On the Howardian Hills, forming the south-western flank of the North York Moors, woodland comprised approximately 8 per cent, 14 per cent and 15 per cent of the land area of the townships of Brandsby, Howsham and Newburgh when mapped in 1746, 1705 and 1605 respectively (Nrth ZQG IV/16; Bev DDX3/15; Nrth ZDV); while at Beadlam and Pockley, lying at the junction of the Vale of Pickering and the North York Moors, it accounted for around 7 per cent of the land area in 1785, most of it found on the steep slopes of the gullies running down the escarpment (Nrth ZEW M 13 and 14a). Woodland was also frequent on the margins of the Pennines and in the Dales cutting through them, although here the varied geology of coal measures, gritstone and carboniferous limestone ensured a rather greater degree of variation, with a mere 1 per cent of Walburn wooded in 1700–43 and only c.2 per cent of Skeeby in 1779, but as much as 27 per cent of Downholme in 1738 (Nrth ZAZ (M) 1; Nrth ZMI; ZBO (M) 1/5). In contrast, on both the higher ground of the moors and in the higher reaches of the vales running into them woodland was everywhere sparse. John Leland noted in the mid-sixteenth century how 'the river sides of Nidde be welle woddied above Knarresburgh for a 2 or 3 miles, and above that to the hedde all the ground is baren for the most part of wood and come, as forest ground ful of lynge, mores and mosses with stony hills … . The principal wood of the forest is decayed' (Woodward 1985, 19). In Wensleydale, similarly, he saw 'very litle wood'. On lower ground, in unenclosed 'champion' townships in the Vales of York and Mowbray, woodland was in even shorter supply. None at all existed in places such as Harlsey (mapped in 1762), Birdforth (1770), Sinderby and Pickhill (1778) and Thormanby (undated, seventeenth century) (Nrth ZNS; Nrth ZDS M2/12; Nrth ZIQ; Nrth ZDS M2/1). Even where woods did exist, as in Healaugh and Catteron or Warmsworth, they usually accounted for less than 2 per cent of the land area (Lds WYL 68/63; Don DD/BW/ E11/7). These variations between wooded and unwooded areas, when averaged out, ensured that overall woodland totals in Yorkshire were low. Tuke in 1800 estimated that in the whole of the North Riding there were only 25,500 acres (10,319 hectares) of woodland – less than 2 per cent of the total land area.

What was true of Yorkshire was broadly true of other northern counties. Woods were concentrated in particular areas, usually away from the high moors and more fertile lowlands, on the sloping ground between the two or bordering major rivers, and especially in places close to industrial areas. Thomas Pennant described the 'thick coppices' near Pennybridge in Lancashire (now Cumbria), 'many of them planted expressly for the use of the furnaces and bloomeries' (Pennant 1776, 33). Indeed, throughout the Lake District the area of coppices seems to have been expanding in the seventeenth and eighteenth centuries, in some cases through the enclosure and planting of common wood-pastures, but it remained nevertheless a region of few woods. Cumberland was described by Culley in 1797 as 'far from being well-wooded', with the largest areas of coppice found on the banks of the river Caldew (Culley 1794, 12). In County Durham woods were likewise largely confined to the banks of major rivers, especially in the area around Durham itself: otherwise 'the face of the county is for the most part naked' (Granger 1794, 47). The first reasonably accurate estimates of woodland area, made in 1895, suggest that all the northern counties had broadly similar proportions of their surface area occupied by woodland: 3.4 per cent in Westmoreland, 3.5 per cent in Lancashire, 3.6 per cent in Cumberland and in Yorkshire, 3.8 per cent in Northumberland and 4.5 per cent in Durham. This, it should be emphasised, was after a period of sustained planting by large estates: the figures may have been significantly lower in the seventeenth and early eighteenth centuries.

Coppiced woodland was thus very unevenly distributed in England in the seventeenth and eighteenth centuries, due largely to factors operating before the period under consideration here. It was relatively sparse in most northern districts and in 'champion' areas, except within the royal forests. It was most abundant in old-enclosed areas, in the south-east of England and the West Midlands, although in only a few counties did it much exceed 10 per cent of the surface area.

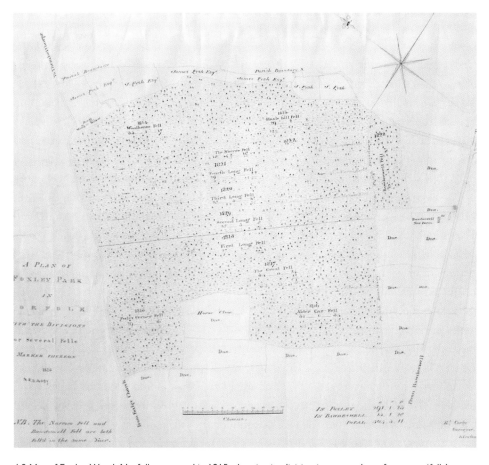

4.2 Map of Foxley Wood, Norfolk, surveyed in 1815, showing its division into a number of separate 'fells', which were coppiced in rotation.

The management of coppiced woods

As described in Chapter 1, traditionally managed woods comprised an understorey of coppice, cut on a rotation to produce a crop of poles, together with a scatter of timber trees. Leases and other documents show that small areas of woodland might be coppiced at one go, but larger woods were divided into compartments called 'fells' or 'falls', usually long-established and with particular names, that were cut in succession. A map of 1815 shows, for example, how Foxley Wood in Norfolk was divided into areas called Foxley Corner Fell, The Great Fell, Alder Carr Fell, First, Second, Third and Fourth Long Fells, Narrow Fell, Woodhouse Fell and Hazle Hill Fell (NRO NRS 4087) (Figure 4.2). Similarly, a map of Canklow Wood in south Yorkshire, mapped in 1810, shows it divided into 11 compartments for felling purposes (Jones 2012, 57–8).

In the Middle Ages woods were, like other demesne assets, usually kept 'in hand', directly exploited by their manorial owners. Wood and timber were felled by estate employees and the income accrued directly to the owner. This sometimes continued to be the case right through the post-medieval period, but on many estates new forms of management were adopted. In a few districts, especially the south-east of England, where woods were numerous and often small, some might be let in their entirety as part of the farm on which they grew, but this was unusual. More usually, the underwood was sold in small portions to numerous different purchasers, each of whom was responsible for felling their own particular allocation. So, for example, an enquiry made in 1594 into Wayland Wood, near Watton in Norfolk, asked whether the wood was sold 'by the acre or otherwise' (NRO WLS IV/6/15), but portions as small as a rood, or even half a rood, are also mentioned in documents. In Hertfordshire this method of management is recorded at Weston and Knebworth as early as the fourteenth century; at Walkern and Broxbourne in the seventeenth century; and at Wain Wood, near Hitchin, on the Lamer estate in Wheathampstead and on the Broxbournebury estate in the eighteenth century (PRO/ TNA SC6/873/21 and SC6/873/25; HALS K100; HALS 9607; HALS B/86; HALS DE/R/ E114/1–10; HALS 27252; HALS DE/Bb/E26). It was still continuing in the early nineteenth century in the woods around Hertford, on the Marden and Panshanger estates and also at nearby Bramfield, where the diarist John Carrington regularly recorded wood sales during the spring months, followed by 'wood feasts' held towards the end of the year, when the purchase money was handed over (Branch Johnson 1973). The feast was held in October: meat and drink were provided by Lord Cowper, which in 1754 cost him a total of 15s. On one occasion the agent had to 'Deduct from the receipts the money which Hale the Woodward was robbed in the night between the 7th and 8th Oct 1813 in returning with it from the Digswell wood feast at Eleston's at digswell water where he had received it from the Purchasers: £130' (Austin 2001). On the Hampton Court estate near Leominster in 1850 the cutting was likewise sold 'by the lug' to a large number of people, who were provided with food and drink at a 'wood day' (HAR A63/111/56/8; A63/111/56/5). Sometimes the price paid for the different portions of the underwood sold in this way varied significantly, as at Walkern in Hertfordshire in 1612, where the price of the parcels ranged from 22s 6d to 50s per rood, and from 9s 9d to 27s 6d per half rood, presumably reflecting the uneven quality of the underwood (HALS 9607). In some woods, however, the precise parcel of underwood to be purchased was determined by drawing lots; the price per lot was fixed, irrespective of variations in the quality of the underwood (Branch Johnson 1973, 46, 84 and 98).

Selling the underwood to a range of individuals was administratively complicated and could damage productivity because the buyers, many with little concern for the next crop, tended to spread the felling and dressing of the wood over a prolonged period of time, thereby hindering the regrowth of the 'spring'. On the estates of Lord Burghley at Hoddesdon in Hertfordshire in the late sixteenth century it was reported that the 473 acres of underwood had been sold, unfelled, to local people for many years in one-acre parcels, each costing £3 (Austin 1996). In 1595 Burghley decided to lease the woods to just two men, John Thorowgood of Amwell and William Keeling from London, for the same price per acre, but local people now paid more for firewood and they petitioned Lord Burghley for redress. In response, the lessees agreed to set aside 60 loads of faggots for sale to the local poor at a penny a faggot, but they were adamant that they were not making excessive profits and that the new arrangement would significantly improve the productivity of the woods (Austin 1996).

Such arrangements, whereby one or more individuals agreed to cut an entire wood for a term of years, were common. Usually the timber was reserved for the landlord, but in some cases the lessee contracted to remove some or all of this as well. The period of the lease might be as short as two years, as at Gressenhall in Norfolk in 1613 (NRO MR 211, 241x 6), as long as 21 years, as in the case of woods at nearby Fransham in the 1540s and 1550s (NRO MS 13253 40B4), or even for 31 years, as at Bressingham in the same county in 1682 (NRO BRA 301/1). Alternatively, the estate might sell the underwood on a yearly basis to a single individual as it came ready for felling, usually by auction to the highest bidder (Jones 2012, 59).

Leases, and to some extent documents generated by estate administration where woods remained 'in hand', reveal much about woodland management in the post-medieval period. The maintenance of adequate boundaries was a perennial concern, lessors often being required to 'make good and strong hedges ditches and Fences … about the Coppice grounds' (HALS DE/P/T262). Problems of trespass and theft are frequently recorded. In 1718 the woods of the Duke of Norfolk around Sheffield and Rotherham are said to have suffered 'great destruction' due to 'being cut downe and Carried Away by Some Idle Disorderly persons' (Jones 2012, 61). Not only was wood taken without permission but, in those woods in which hazel was a major component, nuts were gathered in season, often causing damage to the coppices. The massive banks which generally enclosed early medieval woods do not appear to have been constructed around new woods established after the late sixteenth century, although existing examples were rigorously maintained.

Instead, hedges on smaller banks, little different from those forming the boundaries of fields, became normal.

Poorly maintained boundaries, especially when combined with difficult neighbours, could wreck the coppice. The steward responsible for the woods at Castle Rising in Norfolk reported in 1786 how Mr Beck, 'a very unpleasant tenant', 'Will not keep his cattle out of the wood, though I requested him in the most civil terms to do it. I have now ordered the woodman and Browne to impound the cattle if they are found there, otherwise all the underwood and young wood must be destroyed' (NRO HOW 757/35). The exclusion of grazing stock was not always total, however. Sometimes cattle were allowed into woodland once the regrowth of the coppice had proceeded for several years, as at Tring in the second half of the sixteenth century (PRO/TNA E134/43Eliz/East3). In some cases this was a long-established manorial right freely exercised; in others it could be enjoyed only in return for a payment, as on the Arundel estates around Sheffield in the late seventeenth century (where many of the tenants appear to have abused the agreement by helping themselves to quantities of firewood) (Shf ACM 2/279 and Shf ACM 2/280). Whatever the arrangements, temporary barriers needed to be maintained on the boundaries of newly cut fells to prevent stock in adjacent areas of the wood from wandering in and browsing off the regenerating coppice. In 1545 William Lowyn, custodian of Henry VIII's woods in the manor of Cheshunt, was paid 20 shillings 'for kepyng the hedge & beastes & cattell from the spryng in [a wood called] Rough Cattall [and] for the safe inclosure therof', while Thomas Browne received 13s 4d for 'the kepyng of the hedges & savegarde of Dyvers Coppices' in the king's manor of Bedwell in nearby Essendon (PRO/TNA E315/458). One lease from the Duncombe estate in Yorkshire, from 1708, describes how the lessees were allowed to fell 50 acres of coppice on the condition that sufficient 'garsell' – thorny brushwood such as holly and crab apple – and underwood were left for fencing the cut areas (Nrth: ZEW IV 7/25). It was, however, more common to exclude livestock completely from coppices, and some owners even laid down strict rules for the control of horses used in forestry operations by lessees, one lease for a Norfolk wood stressing the importance of 'preserving the shoots … of such wood from being bitt by the horses fetching the same', while others even refer to the need to keep the horses muzzled (NRO MR 211, 241x 6).

Leases sometimes make detailed provisions regarding routes for carrying and carting out the felled material and some refer to particular aspects of processing it. One lease from Kent laid down that the lessor could dig 'three saw pits in the said two woods provided he does it in such convenient places where there has formerly been coal burnt, or else in some of the ways so there is no damage to the springs' (KHL U120 E6). Saw pits are often referred to, one early seventeenth-century lease from Gressenhall in Norfolk giving the tenants permission to 'dig and make … so many pits called saw pits as shall be … convenient for the sawing, cutting, contriving and converting of the said timber trees' (NRO MR 211, 241×6). Surviving examples are less common than we might expect, and it is notable that a document relating to Redenhall in Norfolk, drawn up in 1737, allowed

4.3 Charcoal burners at work in a wood in Yorkshire in the early twentieth century. Well into the nineteenth century large quantities of coppiced wood were converted *in situ* into charcoal, especially in industrial areas.

the lessors to dig 'saw pitts on convenient parts of the premises if occasion be … provided they fill up and level the same within the time aforesaid' (NRO MC 600/L1–6, 780x9). Charcoal burning is also often mentioned. One lease, from the Arthington Hall estate near Leeds, instructed the lessees to 'take care to make & sett their charcoal pitts & make all their fires that shall be made in the said woods in the most open & waste places therein … so that they nor any of them shall or do receive any prejudice or damage thereby' (Lds WYL160/220/157). In many woods traces of such shallow pits or levelled areas – the base of the charcoal kilns or stacks, covered in turfs, in which the wood was burnt in reducing conditions – survive today (Figure 4.3). In the north of England there appears to have been a particular concern about the effects of the weather on recently cut coppice. The lessees of the wood at Arthington Hall in Yorkshire agreed in 1708 to cut 'sloping … so that the water cannot nor do stay upon the stovens [coppice stools] or roots that are left standing but shall & may slide from them' (Lds WYL160/220/157).

Landowners were particularly concerned to ensure a forward supply of wood and timber, and leases instructed tenants to leave a certain number of trees – often self-seeded specimens – when coppices were felled, usually called 'standells', 'standils', 'weavers' or 'wavers', although other terms were also employed in documents, including 'lordings' and 'blackbarks' in the north. An act of 1543, brought in by a government concerned about the supply of oak to the Royal Navy, stipulated that 12 such successor trees should be left per acre in every wood after felling. Some lease agreements echo this stipulation, such as those drawn up for Gressenhall Wood in Norfolk in 1613 and for Wayland Wood in the same county in 1674 (NRO MR 211, 241x 6; NRO WLS LXIX 25). But even within this county there was much variation, with instructions to leave only eight per acre at Ellingham in 1682 (NRO BRA 301/1) but as many as 20 an acre at Honeypot Wood in Wendling in 1835 (NRO EVL 650/6). These differences hint at variations in the density of timber within particular woods, something confirmed by other kinds of document. The wood at Sporle in mid-Norfolk contained nearly 150 standards per hectare in the 1750s, while at Honeypot Wood, some eight kilometres to the east, there were in 1777 no fewer than 924 trees, all oaks – a density of 92 per hectare – together with 57 pollards, presumably growing on the boundary bank (NRO HNR 135/3/18). In contrast, a survey of Toft Wood in nearby East Dereham, made in 1649, recorded 2,860 timber trees on 143 acres, amounting to only around 40 per hectare (PRO/TNA E178/4988 82482).

As in the wider countryside, most of the timber within coppiced woods was felled when still barely mature. The account books of Clement Chevallier, a landowner in Suffolk, for example, show that of the 138 timber trees taken down in Aspall and Bedingfield Woods between 1728 and 1742 only 17 contained more than 30 cubic feet and only 11 had a quarter-girth of more than a foot, suggesting that most were felled at less than 40 years. The 2,000 oaks recorded in Great Wood on the Toppings Hall estate in Hatfield Peverel in Essex in 1791 had an average estimated volume of only 6 cubic feet (ERO D/DRa E23/6). The 120 trees felled in Stubbs Coppice at Lamberhurst in Kent in 1720 had an average volume of 11.7 cubic feet, while 115 'timber marked to be felled' in woodland at nearby Chatham in 1732 had an average volume of less than 15 cubic feet and the largest contained only 22. The 71 trees sold from the same place in 1737 had an average volume of 19.9 cubic feet and a range of 10–35 cubic feet: the overwhelming majority of trees felled and sold in these woods were less than 20 cubic feet in volume and the majority less than 15 (KHL U120 E6).

This said, woods were more likely than hedges to contain some much larger trees destined for particular purposes, which were more easily grown in woodland, where they were less likely to interfere with crops. In Northamptonshire the proportion of farmland trees on the Boughton estate containing more than 20 cubic feet seldom exceeded 5 per cent, but in The Haw and Ravens Wood in Weekley it was over 9.8 per cent and in Boughton Wood 23.8 per cent. In some cases the presence of large trees raised the average volume of the timber, although usually not by very much. A valuation of woodland made in 1774 at Nunburnholme on the East Yorkshire Wolds described 244 oak and 324 ash, most of which had volumes of less than cubic 10 feet, with the remaining population showing a slow decline in the frequency of volumes up to 50 cubic feet, giving a mean size of around 15 cubic feet (Hull U DDWA/x1/4/3). The same pattern – with a few large trees, but many young ones – is apparent in woods surveyed, or in lists of timber felled, from elsewhere. The 1,773 trees felled in seven unnamed woods on the Easton Lodge estate in Essex around 1810 included a few large trees – one containing as much as 160 cubic feet – but they constituted a small minority. Indeed, only 16 – less than 1 per cent – contained more than 50 cubic feet of wood. Twenty-five (1.4 per cent) contained between 40 and 49 cubic feet and 104 (6 per cent) between 20 and 39. No fewer than 638 trees (36 per cent) were thought to contain between 10 and 20 cubic feet and 990 trees (56 per cent of the total) contained less than ten. Even most of the larger trees, with volumes of 40–50 cubic feet, can hardly have been more than 70 years old. It is noteworthy that a group of oaks in a wood in Coggeshall in Essex in 1787, with average girths of less than 13 cubic feet, were described as having 'done growing' (ERO D/DHt E13).

The evidence of local documents is echoed in the comments made by writers on agriculture and forestry. In the West Riding of Yorkshire, according to Brown in 1799, timber trees in woods were 'supposed to be 60 years old at the time they are ready to go down' (Brown 1799, 82). Contemporaries sometimes saw early felling as a recent development resulting from increased demand for timber or bark, but the documentary evidence suggests it was normal practice and was encouraged by the same kinds of factor that led to the premature demise of most hedgerow timber. Demand for bark certainly rose in the eighteenth century – that stripped from trees felled in Wain Wood near Hitchin in Hertfordshire rose in price from 6d a yard in 1745 to 14d a yard in 1767 (HALS DE/R/E114/1–10). But it had long been a valuable item, as had timber itself. By the seventeenth century demand for the latter was already so great that it might be moved considerable distances. In 1688 Thomas Mills and other landowners in the area around Burgh in Suffolk agreed to sell 107 timber trees to John Haynes, carpenter of St Martin in the Fields, and to John Bennett of Ruislip in Middlesex. The timber was to be delivered by them to a wharf at Woodbridge (Ipswich RO GB 1/5/2/58), suggesting a journey by sea and land of over 150 kilometres.

It is often suggested that the average length of coppice rotations increased from medieval through post-medieval times, from around 7 years in the thirteenth century to as many as 14 by the nineteenth (Rackham 1976, 64–6, 82–3; Rackham 1986, 85, 92; Collins 1989, 484), but the evidence for such a development is not, in reality, very clear cut. In the county of Norfolk, for example, a fifteenth-century survey of Beeston Priory described how one 23-acre wood was divided up into six compartments: five acres were under 7 years' growth, and not valued, but there were '3 acres of 7 years growth, 3 acres of 8 years growth, 2 acres of 9 years growth, 2 acres of 10 years growth and 8 acres of 14 years growth' (Jessop 1887); conversely, at Gillingham in 1717 the underwood was cut every seven years, as it was at Horningtoft as late as the 1830s (TNA IR 18/6019). Either way, most woods in this county, throughout the post-medieval period, were cut at intervals of between 10 and 14 years. An indenture from 1740 concerning Hockering Wood implies a rotation length of 14 years (NRO BER 336 291/7), a figure apparently confirmed by a map of 1805 (NRO 21428 Box F) and a lease agreement for 1828 (NRO BER 336); an agreement for Sporle Wood from 1745 implies 9 years (NRO 20888); another, for Ashwellthorpe in the early eighteenth century, indicates 10 years (NRO KNY 571, 372X3); and one concerning Attleborough Wood, from 1801, stipulates 10 years (NRO MEA 7/4). In Hertfordshire there was also much stability in rotation lengths, at least during post-medieval times. When William Cecil leased his woods in Hoddesdon in Hertfordshire in 1595, 46 or 47 acres were to be felled out of a total of 473½ acres each year, according to the terms of the lease, implying a 10-year rotation. When these and other woods were surveyed in 1785 the dates at which the underwood was last cut were given for 25 woods divided into 64 fells, of which three had last been cut 10 years earlier and three 11 years, but none before this, suggesting little change in rotation length (Austin 1996, 11–18; HALS DE/Bb/E26). At Hadham in 1611 the rotation length was said to be 11 years (HALS 9607); Walker in 1795 thought that most woods in the county were cut on a rotation of around 10 years (Walker 1795, 69); Young in 1804 reported that the woods in the south of the county were cut every 9 or 10 years on average (Young 1804, 145); and at Knebworth a rotation of 11 years is suggested by estate accounts for 1815 (HALS 9607).

Overall, in most counties post-medieval sources suggest that rotations of between 8 and 14 years were usually practised: longer and shorter ones do appear in the documents, but they are unusual. In Suffolk, Young in 1813 described 'cuttings of ten, eleven or twelve year's growth'; at Great Glemham in 1826 the four estate woods were cut on a 9-year rotation (Ipswich RO HD 1000/3), while at Barking in 1831 a wood on the College estate was divided into 11 'fells', suggesting an 11-year rotation (Ipswich RO HA 1/H85/1/2). In Herefordshire it was said in 1852 that 13 years was the normal rotation length, and on the Hampton Court estate in Herefordshire in the 1830s and 1840s the woods seem to have been cut at intervals of between 10 and 14 years (HAR A63/111/56/5: A63/111/56/12). Nevertheless, in particular areas, and for specific economic reasons, rotations might be extended to 18, 20 or more years. This was particularly the case in the north of England and in parts of the west Midlands.

In Yorkshire, for example, the sixteenth century records usually suggest similar rotations to those in the south, as at Settrington in the 1590s, where the woodland was divided into eight coppice springs or 'haggs' that were felled on an 8-year rotation. In some places such rotations were still being practised into the eighteenth century, as on the Blackett Estate near Bradford in 1750, when Garner Wood was found to be of 'perhaps 12 or 14 years growth & ought to be felled with Gill Wood' (Brad 23D98/3/5). But by this time many coppices were being cut on significantly longer rotations, of 20 years or more, as on the Arthington Hall estate in the West Riding in 1708 (Lds WYL160/220/157) or in the Earl of Arundel's woodlands around Sheffield in the early nineteenth century (Lds WYL160/220/157). On the estates of the Marquis of Rockingham in south Yorkshire in 1727 a new management system was adopted by which all the woods were to be coppiced on a 21-year rotation, cutting approximately 40 acres of wood per year (Shef WWM A1273; Jones 2005, 52). Tuke in 1800 suggested that most woods in the North Riding were managed on a rotation of 20 or even 30 years, although he also describes how some were cut more frequently (Tuke 1800, 183–4). In the West Riding, similarly, by the late eighteenth century the coppices were cut at 21-year intervals, according to Brown in 1799 (Brown 1799, 82). Rotations like these were, by the eighteenth century, by no means restricted to Yorkshire, but were common in many northern areas, and in parts of the west Midlands. In Northumberland, for example, it was reported in the 1790s that oak and ash coppices were usually felled on a rotation of between 25 and 30 years (Bailey and Culley 1794, 14). Such long rotations tended to blur the traditional distinction between underwood and timber, which was so important in deciding whether or not the material removed from woods was subject to the payment of tithes – the legal definition of 'timber' (which was usually exempt from tithe) being trees of oak or ash over 20 years of age (Collins 1989, 484). At an enquiry held in Yorkshire in the mid-nineteenth century 'advertisements were produced of the sale of the Estates in which woods of 5, 9, 10, 13 and 15 years growth were called coppice', and 'it was admitted that the woods are generally cut from 28 to 40 years growth', while 'some woods are cut at 23 years growth' (Shf FB/CP/25/11).

The economic contexts of coppicing

These variations in rotation lengths need to be understood in terms of the context of the composition of the underwood (some species grew more quickly than others); in terms of the location of the woods in question (growth tended to be slower in upland areas); and, in particular, in terms of the uses to which poles cut from coppices were put. Like farmland trees and hedges, coppice woods were embedded in very local economies. Economic historians have often emphasised the importance of coppices as a source of firewood, and Collins in particular has discussed how the increasing use of coal as both a domestic and an industrial fuel in the nineteenth century was accompanied by an expansion in the use of coppice wood for new purposes, as the population grew and the economy expanded (Collins 1989, 484–5). But documentary sources leave little doubt that even in the sixteenth and seventeenth centuries much if not most of the material cut from coppices was employed for specialised purposes, and it is noteworthy that Thomas Hale listed the benefits that coppice brought to the farmer in terms of building repairs and the making of 'implements' before those which accrued to 'his Chimney' (Hale 1756, 137).

That is not to say that coppices did not produce wood for fuel. Some of the larger poles were commonly sold as firewood, while smaller material, tied together as faggots, was largely employed as firing. It produced a short hot blaze, and was thus particularly suitable for household baking and brewing, as well as for a variety of industrial processes such as brick- and tile-making. Faggots regularly appear, together with larger pieces of fuelwood, in sixteenth- and seventeenth-century probate inventories. In John Johnson's kitchen in Hertford in 1673, for example, there were 15 loads of 'Faggotts, Blocks and Roundwood' valued at £11 5s; in the woodhouse of William Turner, a gentleman of the same town, in 1683 were three stacks of wood and half a load of faggots worth £2 9s; while in the yard of John Bach, an alderman and gentleman of Hertford, in 1699 there were 'one thousand Jack Faggotts, Six Load Roundwood and blocks' (Adams 1997, 77, 124 and 41). The proportion of material cut from coppice that was used as firewood was probably highest in districts near large cities, although even here it was often mainly the offcuts and waste that went for fuel. Stevenson described in the case of Surrey how 'In the neighbourhood of such a city as

London, and in a county where there is so great a demand for fuel … not the smallest nor the most trifling part of the underwood is useless, or without its value' (Stevenson 1809, 430). In 1724 Defoe described how faggots cut from the woods in north Kent were used in London taverns: 'tis incredible what vast quantities of these are lay'd up at Woolwich, Erith and Dartford; but since taverns in London are come to make coal fires in their upper rooms, that cheat of a trade declines; and 'tho that article would seem to be trifling in itself, 'tis not trifling to observe what an alteration it makes in the value of those woods in Kent, and how many more of them than usual are yearly stubbed up, and the land made fit for the plow' (Defoe 1724, Letter II, 12–13). In these last comments, as so often, Defoe exaggerated: London, continuing its inexorable growth, continued to suck in fuel wood from the surrounding areas well into the nineteenth century; it was still being used by bakers even when coal was widely burned in domestic grates.

Fuelwood was thus an important product, but the wood cut from coppice was used in a vast number of other ways. Scattered widely across England, although especially prominent in particular localities, were a host of minor woodland trades and crafts, including chair-making, the production of crates for glass and pottery, the manufacture of walking sticks and the fabrication of besoms and rakes, all of which might absorb a significant proportion of coppice poles. More important, however, was the production of fencing or hurdles, for which hazel and ash were particularly well suited. Hurdles were employed in vast numbers for folding sheep on the arable fields in areas of light, leached land (Kerridge 1967). John Skayman described in the early sixteenth century how the woods on the Raynham estate in west Norfolk were mainly used to make 'fencing' (Moreton and Rutledge 1997, 115), while in 1730 Edmund Rolfe, the lessee of a large estate in Sedgeford, a short distance away, asked his landlord for permission to plant 40 acres of the worst land on the estate with underwood of sallow, hazel and willow, arguing that such wood was 'being continually wanted by the occupiers of the said estate for hurdles for sheep of which there are great flocks' (NRO DCN 59/30/12). Walker in 1795 implies that much of the material from the woods in Hertfordshire was likewise used to make 'sheep flakes' or hurdles, while Young in 1804 noted the use of 'sallow and willow' cut from the coppices in the south of the county for this purpose, although much of the other wood was made into faggots, many perhaps destined for London (Young 1804, 145). Both ash and hazel were also used for the hoops employed in making barrels, while ash poles were widely used for such things as tool handles and cart wheels. Coppice wood was also employed in the construction of timber-framed buildings – poles of oak, and sometimes ash and elm, were used as minor elements of the frame, while hazel was a major component of the wattle-and-daub infill. Even in buildings constructed of stone or brick, small pieces of wood were needed for minor roofing timbers. In addition, smaller lengths of hazel (and sometimes ash) went to make the 'broaches' used in thatching the roofs of buildings and corn ricks. In 1796 Nathaniel Kent described how the underwood from coppiced woods in Norfolk was used for sheep hurdles, thatching, hoops and general repairs (Kent 1796, 86), and the wood accounts for the Henham estate in Suffolk from the

1820s refer to poles, long hoops, short hoops, thatching spars and broaches, 'sways', stakes, rake handles and hurdles. Here, perhaps typically, only the residue appears to have been destined for fuel, the smaller brush faggots being sold to bakers and cottagers for 'oven wood' and the off-cuts for cottage firing. In short, most woods produced some firewood, but coppices mainly provided high-quality poles that were used in a range of ways in the local economy or were converted *in situ* into specialised commodities such as charcoal.

Much wood was converted into charcoal, even in remote rural areas. A lease for the woods at Gressenhall in Norfolk, drawn up in 1613, for example, allowed the lessees 'to make and sett within the said wood so many harthes called coal harthes' as they needed to convert as much of the underwood as they wished into charcoal (NRO MR 211, 241x 6). In the early seventeenth century Sir Arthur Capel was paying about £12 18s each year to have charcoal made in his woods at Walkern in Hertfordshire in order to heat Hadham Hall (HALS 9607). The amounts of charcoal produced were, however, significantly larger where woods lay close to large cities, especially London. Charcoal is lighter than firewood and has a higher calorific value, and could thus be transported economically over longer distances; Young described how in the Weald of Sussex the underwood was 'converted into hop-poles, hoops and cordwood; the principal part of the latter goes to London in the shape of charcoal' (Young 1808, 471). And charcoal was also used for a variety of industrial processes. It was a raw material in gunpowder production and the main fuel employed in iron smelting, albeit gradually giving way to coke in the course of the eighteenth century (Hammersley 1973). In Ecclesall Woods on the outskirts of Sheffield the remains of nearly 200 hearths have been linked to the industry (Gowans and Pouncett 2003). At Hornby in north Lancashire it was said in 1757 that coppice wood sold at a high price 'on account of the fact that there were several iron furnaces within twenty miles of these woods; the uses they make of it is to Coal it or make it into charcoal for the iron furnaces' (Holt 1999, 17). The close connection between industrial production and woodland management is evident from the fact that forges frequently owned or leased areas of woodland. In the 1650s, for example, Lionel Copley, the most important ironmaster in south Yorkshire, was leasing woods from numerous different landowners, including 13 belonging to the second Earl of Strafford at Wentworth Woodhouse (Jones 2012, 57).

Industrial production shaped local woodland management in other ways: as already emphasised, industry did not 'use up' woodland but ensured its continuing economic value, if correctly managed (Jones 1993; 1997; 2005; Hammersley 1973) (Figure 4.4). Whitecoal – that is, the rapidly dried wood widely used for smelting lead before the late eighteenth century – was made in many of the woods located in lead-mining districts, with over 100 of the distinctive 'Q-pits' (pits with an external bank featuring a downslope spout or gap) associated with its production being recorded from the Ecclesall Woods near Sheffield alone (in addition to the charcoal pits already noted). In 1657 the woods were leased for the production of 'charcole and whitecole' to a 'lead merchant', who was permitted to make 'pits and kilns for the coaleing of the same' (Jones 2012, 79). Above all, timber and

4.4 William Williams' *Afternoon View of Coalbrookdale* (Shropshire) of 1777, showing smoke rising from furnaces set in a wooded environment. Early industry, as this illustration vividly shows, did not 'use up' woodland, but instead encouraged its retention and careful management.

larger coppice poles were in high demand for pit-props for coal or other mines, as writers such as Brown and Bailey emphasised (Brown 1799, 127). It is largely for these reasons that rotations were generally longer in places in or close to mining districts. Plymley thus described how in Shropshire:

> Underwoods are very extensive; they consist chiefly of oak … . Large quantities of oak poles are used for different purposes in the coal-pits; as they are required to have some strength, they are seldom fallen before 24 years growth, and the bark (used in tanning leather) is an object of great importance … . (Plymley 1803, 219)

Yet industry encouraged long rotations in other ways. The charcoal used for smelting was best made from large poles, especially of oak, and the best bark for tanning was stripped from poles aged around 20 years (Jones 2012, 80). On the Milford Estate near Leeds in the eighteenth century bark accounted for around 20 per cent of the sale value of oak trees, but at Hutton Rudby in the 1630s the figure was as high as 33 per cent (Lds: WYL500/939; WYL 100/EA/13/38), significantly above the values recorded in Midland and southern counties, suggesting both a greater demand for bark in this more industrialised area and limited supplies in what was generally a poorly wooded region. Given the scale and variety of demand in such districts, however, it is not surprising that different woods on a property were sometimes managed under differing regimes. On the Duke of Leeds' estate at Kiveton Park, for example, there was a mixture of shorter and longer felling cycles, which produced a wide range of pole sizes suitable for 'hop poles, scaffold poles, cordwood, pit wood ("puncheons"), heft wood, hazel hoops, hedge bindings' (Jones 2005, 52).

New forms of industrial production influenced coppice rotations in other areas. Hops for the brewing industry began to be planted on a significant scale in England in the sixteenth century, and by the seventeenth century the woods of Surrey, Sussex and Kent were already producing vast numbers of the poles required to construct the frames on which the hop plants grew. As early as 1654 14,000 'hoppoles' were cut from one wood on the Sackville Estate in Kent (KHL U269 E27), and by the early nineteenth century they were 'the chief article which make woods valuable' in that county (Boys 1805, 137). Hops were also widely cultivated in Herefordshire and the south of Shropshire, again providing an important market for coppice poles, as on the Hampton Court estate in the early nineteenth century (HAR A63/111/56/5). The poles needed to be long but sturdy and, where they were a major product, the normal local rotation of between 10 and 14 years was extended to as much as 18 years.

The management of coppices was thus firmly embedded in the character of the local economy. But the nature of local markets could change through the seventeenth and eighteenth centuries, as patterns of industrial production and forms of technology evolved and developed, with or without a significant impact on the extent or management of woodland. Arthur Young described how, as coke replaced wood as the main fuel used for iron smelting, 'the iron works which took off and consumed such quantities of wood deserted the Weald': yet, contrary to expectations, this did not mean that the price of wood declined in the locality. On the contrary:

> Such a new demand has been created for the consumption of these extensive underwoods, in burning limestone for manure, and the great and still increasing demand for hop poles; all this, with an increased population and a better system of husbandry, which everywhere pervades the whole country, are the reasons why woodlands have been rising in value. (Young 1808, 169)

Wood-pastures

Historians and historical ecologists have devoted much attention to coppiced woodland, rather less to wood-pastures, although in particular localities these could occupy, well into the period studied here, a similar area of ground. It is often assumed, or implied, that by this time grazed woodlands were in full retreat, gradually degenerating to more open ground. Their trees were vulnerable to damage from stock, through the stripping of bark, for example, or the compaction of the ground above their root systems. When trees were felled, died or blown down by the wind, moreover, it was difficult to replace them because of the intensity of grazing, especially on commons, where use-rights were exercised by large groups of people (Rackham 1986, 121–2). In fact, although it is difficult to provide precise figures – not only because of the patchy nature of the evidence but also, as already noted, owing to problems of defining the boundary between 'pasture' and 'wood-pasture' – there is no doubt that the speed with which wood-pastures declined has often been exaggerated.

In the Middle Ages extensive tracts of wooded common survived within many royal forests, where they lay interspersed with areas of embanked coppice. Forests were, as we have seen, particularly important in Midland 'champion' districts, where they formed large islands of wooded ground within otherwise rather open landscapes, but many examples could also be found in old-enclosed areas, especially in the west (Figure 4.1, above). The history of forests in the post-medieval period is both poorly understood and, in some ways, inadequately studied. The Crown gradually came to use them less and less for hunting, and as many as 47 examples were alienated altogether during the sixteenth and seventeenth centuries, mainly passing into the hands of wealthy courtiers, including Neroche in Somerset, Bernwood in Buckinghamshire and Hatfield in Essex (Langton 2005, 4) (Figure 4.5). Others were partially alienated. In Northamptonshire, for example, the

4.5 Old hornbeam pollards in Hatfield Forest, Essex, with embanked coppices in the background. Hatfield, near Stansted airport, is probably the best-preserved forest landscape in England, although this section of wood-pasture came into existence – converted from enclosed coppices – only at the end of the seventeenth century.

royal parks in the three forests of Salcey, Whittlewood and Rockingham were disparked and the woodlands in Rockingham progressively sold or granted away, while Geddington Chase in Rockingham was alienated in its entirety in 1676 (Pettit 1968). New owners of forests usually maintained them in their traditional state, however, and often continued to use them for hunting, something reflected in the fact that laws continued to be passed – including the famous Black Act of 1722 – to protect deer in 'forest, chase, purlieu, paddock, park or other ground where deer are or have been normally kept' (13 CarI C10).

Yet while monarchs used forests less and less for hunting, those remaining in the hands of the Crown were by no means neglected, but instead regarded in new ways – as a source of income and materials (James 1990, 123). Substantial sums might be raised by leasing the coppices, often – as in the cases of the Forest of Dean or the great forests of Sussex and Kent – to local ironmasters. Money was also made, especially under the early Stuarts, through the more rigorous enforcement of forest law. Above all, as the demand for ship timber escalated steadily through the seventeenth and eighteenth centuries (James 1990, 119–24), forests were seen as a source of timber for the Royal Navy, especially those examples – such as the New Forest – lying near to the sea, to navigable waterways or to major dockyards. The scale of such essentially practical and financial interests is again reflected in the volume of legislation. Between 1660 and 1850 around 450 laws primarily relating to forests were passed, including the 1698 act 'for the increase and Preservation of Timber in the New Forest', which allowed the Crown to enclose substantial tracts for timber planting (Paley 2005, 30; James 1990). The scale of government interest is also clear from the number of surveys and enquiries conducted, most notably those made between 1787 and 1793, following the passing of the Crown Lands Revenue Act of 1786.

But forests represented a challenge to the state. Their ancient and complex organisational structures, the various measures to protect the deer and the overlapping rights exercised by communities living in and around them all militated against sustainable management. In particular, local people did all they could to subvert attempts to preserve wood and timber. In many forests in the Middle Ages the commoners could turn only cattle (in the modern sense) into the forests, but by the sixteenth century sheep, more damaging to trees and underwood, were commonly being pastured on a large scale. The coppices were inadequately fenced and often grazed out of season, and overstocking was endemic: when Rockingham in Northamptonshire was enclosed many commoners claimed the right to graze unlimited numbers of animals 'all year round' (NHRO Brooke of Oakley 318/1). The forest officials were often as guilty of despoiling the woods as the commoners, claiming the right to lop trees (ostensibly to provide browse for the deer) and in some cases the right to graze in the coppices. In 1720 the steward of the Boughton estate in Northamptonshire, much of which lay in Rockingham Forest, informed Lord Montagu that 'horses were put into copses of seven years old. I have seen them and can find any manner of damage done ... they say it is a privilege that the keepers of this and other chases have' (Toseland 2013, 42). At Cranbourne Chase in Wiltshire in 1791 'The damage done to the woods … is very considerable, so much so that the underwood is in very few instances fit to cut under 18 years, which would otherwise be as fit at 12 and in some instances at 9' (Cheeseman 2005, 69). The right enjoyed by commoners to collect 'sere and broken' wood, theoretically fallen material, was often interpreted to include green and growing wood, either from trees or from coppices. In the Forest of Dean commoners assisted nature in providing 'sere' wood by deliberately barking healthy timber trees (Hart 1966, 192). At Wakefield in Northamptonshire in 1623 seven trees were stolen, supposedly for maypoles (Pettit 1968, 125). The right to cut oaks for 'coronation poles' was similarly claimed at Cliffe in Rockingham in 1702 and in the forests of Whittlewood and Salcey (all in Northamptonshire) in 1714 and 1727, when troops had to be called in to control the situation. Attempts were made to preserve timber on the forest plains by leaving 'seed' trees – large oaks or ashes that could drop acorns or keys onto the surrounding ground – but such initiatives were often thwarted by the intensity of grazing.

Indeed, forests were often areas of social tension and unrest. The opportunities they provided for eking out a living attracted squatters and caused local over-population, and there were recurrent disputes, often violent, not only between forest officials and local people but also between different groups among the latter. There were numerous clashes between villages concerning grazing and other rights or, as in the Forest of Dean, between commoners and the ironmasters who leased the coppices (Hart 1966, 95–104). Deer were poached on a large scale, and were continually disturbed during the fawning season by people entering the woods to collect wood and nuts. In Whittlewood in the eighteenth century fights between forest officers and 'offending nutters' were said to be as fierce as any with poachers (Linnell 1932, 21, 103). In May 1659 'divers people in a tumultuous way in [the forest of] Dean did break down the fences and carry away the gates of certain coppices enclosed for the promotion of timber, turned in their cattle and set divers places of the Forest on fire, to the great destruction of the young growing wood' (Hart 1966, 149). In more general terms, forests were viewed as an anachronistic anomaly, a hangover from the Middle Ages, that sat uneasily with modern concepts of common law. Forests, in short, were a byword for lawlessness and poor management. Their very existence seemed an offence to improvement. Most educated people believed that they should be enclosed and used for growing timber or corn, and would have agreed with comments such as those made by James Donaldson about Whittlewood on the Northamptonshire/Buckinghamshire border in 1794: 'I know of no land in England, of equal staple, worse misapplied than a great part of this forest' (Donaldson 1794, 132–3).

Yet all this does not necessarily mean that the forest plains were significantly denuded of trees. Some certainly grew more open over time. The plains of Weldon, Benefield and Deenethorpe in Rockingham Forest were almost devoid of trees in 1580, but the density of charcoal hearths recovered by archaeological survey strongly suggests that they had been wooded in the Middle Ages (NHRO Westmorland 4 xvi 5 and 30; Watson Mun. A.5.22,D2817). On the other hand, a survey made in 1565 of the three Northamptonshire forests (Whittlewood, Salcey and Rockingham) recorded 93,942 oak standards (valued at £46,355) in the plains and in the enclosed woodland still in the hands of the Crown, and a further 14,198 (£4,609) in the remaining royal parks: a by no means negligible figure, given that the land in question probably amounted to some 200 square kilometres (a density of around five trees per hectare), and that it excludes pollards (PRO/TNA LRRO 5139). On the wood-pasture commons of private chases, especially in the north, much woodland often remained at least into the later seventeenth century. In 1650 Loxley Chase in south Yorkshire was described as 'one Greate wood called Loxley, the herbage consisteth of great Oake timber' (Jones 2012, 48). Even in the following century many forest plains remained full of trees. As late as 1790 it was still said that there were 'great Quantityes' of 'Pollards and decayed trees' in Rockingham Forest in Northamptonshire. 'Thousands' of trees in Whittlewood Forest were regularly lopped by the keepers (*Journal of the House of Commons* 1792). Timber perhaps survived best where areas of forest had been brought into private ownership, even though common rights were still being exercised. In 1749 no fewer than 20,664 trees were recorded in Geddington Chase in Rockingham (Boughton House archives). But even where they were still Crown property many forest plains remained well treed, and even well timbered. As late as 1807 Rudge was able to describe how 'notwithstanding the constant depredations committed, there still remains a large quantity of usable timber' in the Forest of Dean. In 1788 the forest contained 46,000 substantial oaks, 'besides unsound trees, which are numerous, and a considerable quantity of fine large beech, and young growing trees'. Rudge believed that the amount of timber had been reduced by more than two-thirds in the course of the eighteenth century, but 1,000 loads were still being felled each year for the naval dockyards (Rudge 1807, 249, 252). Arthur Young similarly noted 'a considerable number' of thriving trees on the plains of Wychwood Forest in Oxfordshire, alongside the pollards producing 'brush-fuel and browse for the deer'. His concern, like that of many contemporaries, was that more timber would be produced in enclosures, where trees could be preserved from lopping and the effects of grazing: 'when these trees are compared with the space of land in which they are found, they cease to be objects of any consideration' (Young 1809, 327–8). In other words, the parlous state of the royal forests suggested by some contemporary writers needs to be viewed in context and taken with a pinch of salt. Over time their wood-pastures did indeed gradually degenerate, but large numbers of pollards and even timber trees remained, as is clear from those rare cases where, following alienation or enclosure, areas of forest were retained, either to provide recreation, through inertia, or to indulge the antiquarian or romantic interests of owners (Figure 4.6). It was usually formal enclosure and disafforestation in the nineteenth century, rather than gradual degeneration, that destroyed the great forest wood-pastures.

Historical attention has tended to focus on forests, and less research has been undertaken into the many other tracts of common wood-pasture that still existed in England at the end of the Middle Ages – on the poorest soils and in remote locations,

4.6 Old beech pollards, closely spaced, within Epping Forest, Essex. These magnificent trees give a good indication of how densely wooded some commons, especially in royal forests, remained into the nineteenth century.

and especially on high watersheds. In south-east England, for example, extensive wooded commons existed in the Weald of Kent and Sussex, on the crest of the Chiltern Hills in west Hertfordshire, Buckinghamshire and Oxfordshire and on the poor pebble gravels and sands of the Bagshot and Claygate Beds in south Hertfordshire, south Essex, Surrey and Kent. Extensive common woods also existed in many northern areas, especially on the fringes of the principal upland masses, well into the seventeenth century. Wherever they were located, they faced mounting threats. Firstly, manorial lords continued – where they could – to enclose portions, converting grazed woodland to coppice or, by the end of the eighteenth century, plantation, something made easier by a parliamentary act of 1756 that allowed partial enclosure for this purpose provided the principal tenants were in agreement. This phenomenon may have been particularly frequent in northern areas, especially in and around the Lake District, owing to the scale of industrial demand. At Hornby in north Lancashire, for example, an agreement drawn up in 1757 describes recent rent abatements made to one tenant 'for his want of grass of Barkin Woods … now preserved for the sale of young spring' (Holt 1999, 17). Secondly – and more importantly – as with the forest plains, poor regulation and over-exploitation often led to steady degeneration, over time, to open ground. In the late seventeenth century Barnet Common in south Hertfordshire was said to have been 'formerly a wood, but had been of recent years laid waste, and used as a common' (Page 1908, 329–37). In 1584 'divers parcels of wood' were felled and sold for £160 from the 200-acre common 'woodgrounde' called Mayne Wood in the parishes of Tring and Wigginton in the Chilterns, but by the beginning of the seventeenth century the common was being described as the 'waste or pasture called The Mayne Wood' (PRO/TNA E 134/4Jas1/Mich19) and by 1650 wood and timber growing on *all* the extensive waste of the manor of Tring was valued at just £10, and the customary tenants gathered only small shrubs and bushes from within what was still being described as West Wood (PRO/TNA E317/Herts/29). As the trees disappeared many of the Chiltern commons became heaths, characterised by large tracts of gorse, heather and broom. By the mid-seventeenth century the southern end of nearby Berkhamsted Frith, wooded in the sixteenth century, had lost its tree cover and was marked on a map of 1638 as 'Barkhamsteed Heathe with out trees' (HALS 1985). In 1748 Pehr Kalm, the ever-observant Swedish visitor, described how much of Berkhamsted common was overgrown with gorse, 'not much over a hand's breadth high because the poor people are continually cutting it down to the ground and taking it home as fuel' (Lucas 1892, 25).

The late emergence of heathland in these areas is an important reminder not so much of the fact that heaths are artificial environments that were created through the over-exploitation of woodland but that they could emerge at a remarkably late date. Not all open heaths, in other words, had prehistoric or early medieval origins. Even in East Anglia, where some of the greatest areas of heathland were to be found, many examples (on the more gravelly soils especially) only gradually degenerated to open land through the sixteenth and seventeenth centuries. In the 1590s the inhabitants of Marsham in Norfolk accused James Brampton of having, among other misdemeanours, 'felleth downe woode growinge uppon the common contrarye to the custome of the mannor' (Smith *et al.* 1982, 242–3); and, when first mapped in the late eighteenth century, the parish consisted only of arable and heath. What were later treeless heaths are shown as at least partly wooded on a number of early maps from the same county, including surveys of New Buckenham (1597) (NRO MC 22/11), Haveringland (1600) (NRO MS4521), Castle Rising (early sixteenth century) (NRO BL71) and Appleton (1596) (NRO BRA 2524/6).

Yet, outside forests as much as within them, we should not exaggerate the speed with which common woods degenerated to open pasture. Negative comments by contemporaries were often related to the absence of *timber* in such locations, rather than to a paucity of trees *per se*. The parliamentary enquiry into timber supplies carried out in 1790 included a question about whether there was much timber surviving in the 'commonable woods' found in various areas, and one respondent suggested that 'In the Countries where poor People abound, and Woods are near, and Fuel dear, great Waste is made by bowing and cutting off the Branches and the Limbs, to the utter Destruction of the Timber, especially when young' (*Journal of the House of Commons* 1792, 293). Another noted that 'the Quantity of Timber in such Woods or Commons is inconsiderable, but in general headed, and made pollards'. He added that he did not believe that, by this time, there was 'much

Commonable Wood of considerable extent'. But this may have been a relatively recent development, and an exaggeration, for contemporary maps, surveys and descriptions often indicate that many of the largest commons still retained significant numbers of trees.

The conjoined Surrey commons of Epsom, Leatherhead and Ashted – 12,000 acres (c.4,850 hectares) in all – were said, as late as 1794, to contain 'large numbers of hornbeam and other pollards', while around a fifth of Wimbledon and Putney commons, on the very doorstep of London, was still occupied by oak and hornbeam pollards (Stevenson 1809, 467–8). On Childerditch Common in Essex in 1765 there were thought to be 1,993 pollard trees as well as 2,384 'spires' or young trees (ERO D/DP E20/1). In the same county, a detailed survey of 1722 shows a total of 29 mature timber trees and 321 pollards growing 'in the hedgerows' of Drakeshill Farm, Navestock, and a further nine timber trees and 83 pollards free-standing 'within the several enclosures'. But on the adjoining Slade Common the farm had rights to 68 timber trees and to no fewer than 891 pollards: presumably others, associated with other farms, grew elsewhere on the common (see below, Figure 5.1) (ERO D/DU 583/2). Slightly earlier, in 1695, Cheshunt Common in south Hertfordshire, covering 1,186 acres (c.480 hectares), reportedly contained 24,000 hornbeam pollards, a density of nearly 50 per hectare (HALS 10996 A/B). An average of around 3,800 of these were lopped each year over the 20-year period from 1658 to 1678, suggesting a short pollarding cycle of around six or seven years (HALS B/86). On Broxbourne Common, a short distance away, which covered only around 60 acres, there were at least 4,420 pollards in 1682, an apparent density of 180 per hectare (HALS B/86). In 1748 the Swedish visitor Pehr Kalm described how the great tract of undulating common land that existed 'between Cheshunt and Bell Bar', to the south of Hatfield in Hertfordshire, 'was covered with tufts of ling [heather], between which bracken flourished and swamps abounded. But there was scarcely any grass. Sheep grazed here. In places *Carpinus* (hornbeam) grew fairly densely to a height of six feet, and the tops of it were cut for fuel' (Lucas 1892). Even small 'greens', of the kind common in old-enclosed clayland countryside, might continue to carry significant numbers of trees into the eighteenth or even nineteenth century. The pollards sold from Tingate Farm, Broxted, Essex, as late as 1852 included 60 'on the green' as well as 17 'in the lane' and 26 described as 'road side' (ERO D/F 35/3/135).

The extent to which pollards survived on commons was the consequence of the complex interplay of two principal factors: the scale of local demand for wood and the strength of organisational structures that could control exploitation and limit the effects of grazing. It is clear that in many places customary mechanisms existed that allowed tenants to protect new planting on commons (Dallas 2010). During a legal dispute in the late sixteenth century concerning the commons at Pulham in south Norfolk, for example, it was stated that 'The tenantes of the said manor have used to make benefitt of the trees growing upon the common near their houses which were planted by themselves and their predecessors' (NRO NAS II/17). A survey of the manor of Gressenhall in the same county, drawn up in

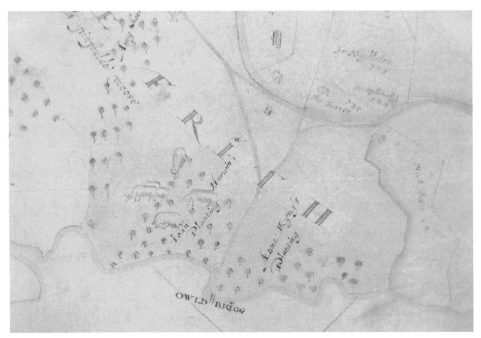

4.7 Map of Gressenhall in Norfolk, surveyed by Thomas Waterman in 1624, showing a partially wooded common. The various areas of trees are labelled as 'plantings', and are allocated to particular named tenants.

4.8 Pollarded oaks on the edge of Fritton Common in south Norfolk. The trees grow in straight lines, showing that they have been deliberately planted.

1579, describes how tenants admitted to holdings received one or more 'plantings' (NRO MR61 241X1), and a map of 1624 shows that these were wooded areas situated on the various commons of the parish, each of which was associated by name with the owner close to whose home it was located (NRO Hayes and Storr 72) (Figure 4.7). The historian Francis Blomefield in 1739 similarly reported how the tenants of his home parish of Fersfield, also in Norfolk, had 'liberty to cut down timber on their copyholds, without licence and also to plant and cut down all manner of wood and timber on all the commons and wastes against their own lands, by the name of an outrun' (Blomefield 1805, Vol. 1, 739, 95).

Blomefield describes similar customs at other places in the south of the county, including Kenninghall, Diss and Garboldisham (Blomefield 1805, Vol. 1, 220, 263). Once again the trees established by the commoners were close to their 'own lands', and in this context it is noteworthy that where old pollards survive on commons they tend to be concentrated towards their margins. The old pollards growing on the western side of Fritton Common in Norfolk, for example, appear to be arranged in straight lines, clearly indicating that they were deliberately planted (Figure 4.8). Not surprisingly, enclosure awards often include claims made by commoners for compensation for trees planted on commons, which would be lost following enclosure. Manorial lords also made such claims, and at the enclosure of Shipdham in mid-Norfolk in 1807 the Earl of Leicester made a submission for the value of:

> All trees, and all bushes and thorns planted or set by him or his predecessors, or his or their tenants, upon the said commons and waste grounds, contiguous or near to any of his said messuages or farms, which have been usually lopped, topped, pruned, or cut by him or his predecessors, or his or their tenants. (NRO BR90/14/2, p. 17)

Such customary systems seem to have operated in many districts, not just East Anglia, significantly retarding the degradation of common woods to open pasture.

In some areas, especially in the vicinity of London, manorial lords appear to have taken a particularly active role in the management, and perhaps the preservation, of trees on commons. They had a strong interest in so doing because, in districts such as south Hertfordshire, they seem to have managed to assert their ownership of the lops of the pollards as well as of the bollings. When Lord Burghley leased his Hoddesdon woodlands to John Thorowgood and William Keeling in 1595, for example, the grant included 'all those lops of trees and bushes growing … in or upon Goodsgreen and Redhillstreet' (HHA Deeds 198/38). Over the 20 years between 1658 and 1678 the lord of Cheshunt manor made an average of £75 per annum from the pollards on Cheshunt Common. The wood was sold uncut at a set price, usually in batches of 100 or 50 heads, to local residents who lopped them themselves (Rowe 2015, 310–12). It is noteworthy that these commons are still shown with extensive tree cover on Dury and Andrew's map of Hertfordshire, published in 1766. Such careful retention of manorial control presumably reflects the proximity of London, the market it provided for firewood and charcoal and thus the inflated prices locally for wood, and Anne Rowe has recently suggested that pollards of hornbeam were being actively planted on Enfield Chase, Northaw Common and other Middlesex and south Hertfordshire commons in the late seventeenth century, as a way of maximising the income from the manorial 'wastes' (Rowe 2015). Lordly planting of commons is suggested elsewhere. The remarkable beech pollards found in and immediately around Felbrigg Great Wood in north Norfolk may represent not (as has been suggested) the survivors of an isolated pocket of native beeches, the most northerly indigenous examples of this species to be found in England (Rackham 1976, 27; Rackham 1986, 141), but instead trees planted on heathland by William Windham of Felbrigg Hall, who undertook a sustained forestry campaign in the area from c.1676. None of the surviving specimens appear to pre-date the late seventeenth century, and nineteenth-century writers such as James Grigor, while describing many old and large trees at Felbrigg, fail to mention any ancient beeches there (Barnes and Williamson 2011, 110–12; Grigor 1841).

Common wood-pastures thus disappeared less rapidly than we might think. Not only in royal forests, but also more generally, a significant number of commons retained respectable numbers of pollards and even timber trees well into the eighteenth or even the nineteenth century. And, in addition, many private wood-pastures, in the form of deer parks, continued to exist. In the course of the post-medieval period parks became more closely associated with major residences and grew more open and more ornamental in character. Yet this was a gradual process and for a long time most remained quite densely wooded. A survey of Hatfield Middle Wood, made in 1669, listed no fewer than 1,685 pollards of oak and beech, but in addition gave a large number of values for groups of hornbeam pollards, each said to represent 40 trees, suggesting that there were, in all, nearly 10,000 pollards growing on the park's 350 acres, a density of around 70 per hectare (Austin 2013). Blickling Park in north-east Norfolk was established in the 1560s and expanded in the 1740s; in 1756 a survey valued the mature trees in the park at no less than £2,780, a huge sum, constituting around a third of the total value of timber on the estate (NRO MC3/252). A supplementary examination added a further £120 worth of timber in the park, including 70 'firs' (probably Scots pines), 75 horse chestnuts, 107 beech, 200 'small oaks', 160 oak pollards and 10 lime pollards. Both valuations explicitly ignored the 3,780 'Young planted trees in the Park' with girths of less than six inches, as well as those in 'near 30 of new Plantations'. Almost certainly the older pollards, without timber value, were also excluded, as they were from the survey of the rest of the estate. As the taste for more manicured parkland developed through the middle and later decades of the eighteenth century, under the influence of Capability Brown and his contemporaries, parks did finally become less densely treed, and pollarding within them declined. Educated opinion, as we shall see, turned overwhelmingly against the practice. This said, managed wood-pastures could still be found in many parks. The trees growing in the 320-acre park at

4.9 Ancient oak wood-pasture at Calke Abbey Park, Derbyshire. Although old pollards are a feature of farmland in northern districts, the most important concentrations tend to be found in wood-pastures, especially old deer parks.

Broxbournebury, Hertfordshire as late as 1784, for example, included 381 oaks, 172 elms, 115 chestnuts, 20 walnut, 8 ash and 3 beech, plus 1,325 young trees (spars) and no fewer than 991 pollards (HALS DE/Bb/E27). At places such as Chatsworth and Calke Abbey in Derbyshire, and Moccas in Herefordshire, parkland wood-pastures have survived to this day, albeit no longer intensively managed (Figure 4.9).

There is some evidence that the trees in parks were often retained for longer than those growing in woods or hedgerows. Many of the most ancient trees surviving in England today are found within parks and former deer parks, as at Chatsworth in Derbyshire or Windsor Great Park, as were many of those recorded in the past, but which have since died or been removed, such as the huge Winfarthing Oak in Norfolk. Early commentators often singled out the size and antiquity of parkland trees. Evelyn described the huge oaks in Sheffield Park in Yorkshire, one of which was allegedly so big that, when felled, men on horseback on either side of its trunk could not see each other (Evelyn 1664, 229). Examples of both timber trees and pollards were presumably retained in parks because, while the latter had important economic functions, they were primarily a recreational resource and a symbol of status and were valued for their beauty, antiquity and associations.

Wood-pastures, especially on common land, were thus a significant feature of the landscape in many parts of England well into the post-medieval period, and their almost complete disappearance since has ensured that we now think of 'semi-natural woodland' largely in terms of coppices. In Hertfordshire, for example, careful analysis of Dury and Andrews' county map of 1766, together with near contemporary local maps, suggests that while coppiced woodland accounted for around 6 per cent of the country's land area, wood-pastures on commons and in parks made up at least another 1.5 per cent. By that time wood-pastures were, it is true, often degenerating to open pasture under the pressure of exploitation, and through the decline of ancient forms of manorial and communal management. But in many cases it was only parliamentary enclosure and disafforestation in the nineteenth century that saw their final demise. When the Ordnance Survey draft 2-inch map was being surveyed in south Hertfordshire in 1805 the surveyors arrived at the great wooded common of Northaw halfway through its enclosure and dutifully marked on the map its anticipated fate: 'clearing for enclosure'. The handful of wooded commons that remain in the country are rare and precious survivors.

Conclusion

In the seventeenth and eighteenth centuries, as in earlier periods, coppiced woods were intensively managed. Far from being 'natural' environments, they were, in effect, factories for the production of wood and timber. They made some contribution to the country's insatiable appetite for fuel, but their role in this respect is perhaps exaggerated. Apart from the fact that they generally occupied a small proportion of the land surface, documentary sources make it clear that they were also, and in many cases primarily, a source of the raw materials required for building, farming and industry. Their management thus varied with the character of the local economy, and this had probably been true for centuries. Wood-pastures, which at the start of the seventeenth century probably occupied at least as much land as coppices, were more important as a fuel source. By 1600 they were already in decline, but their demise was gradual and a significant proportion still survived when forests and most commons were enclosed in the early nineteenth century. These, too, were far from being natural environments, the tattered remnants of Vera's primeval savannahs. Many, perhaps most, were sustained by deliberate planting. All this raises in turn the question of how far the species composition of these various forms of woodland – and, indeed, of the trees found in the wider countryside – was the consequence of economics and of human choices, rather than being the outcome of primarily 'natural' processes.

CHAPTER FIVE

The nature of trees

Oak, ash and elm

The most striking characteristic of tree populations in England in the seventeenth and eighteenth centuries was the overwhelming dominance of oak, ash and elm. Surveys and maps almost invariably show that these together accounted for between 85 and 100 per cent of the trees growing on farmland, and in most cases all of the trees (as opposed to underwood) found in enclosed, coppiced woods. To take examples from a single county, Suffolk: oak, ash and elm, in various combinations, together comprised 87 per cent of the trees on a farm at Campsea Ashe in 1807 (Ipswich RO HD 11:475), 92 per cent of those growing on the lands of Thurston Hall Farm at Hawkedon in 1823–8 (Bury St Edmunds RO HA535/5/35) and 93 per cent on a property at Chevington in 1820 (Bury St Edmunds RO HA 507/2/460), while at Badwell Ash in 1730 no other species at all were recorded (Bury St Edmunds RO 613/642/1). The dominance of the countryside by these three species was not total, of course. Wood-pastures, whether private or on common land, sometimes boasted a significantly different range of trees, as we shall see, as did areas of marsh or fen. Some limited districts, for environmental or economic reasons, might deviate from the norm, and even single farms might display an idiosyncratic range of trees. On Drakeshill Farm in Navestock, Essex, in 1722, for example, oak, ash and elm made up only 39 per cent of the mature trees present (Figure 5.1) (ERO D/DU 583/2). There is a marked contrast in this case, however, between the timber trees and the pollards. The timber trees on the farm were all oak, ash and elm; it was the diversity of the *pollards* (featuring maple, lime, alder, hornbeam, walnut and even wild service) that accounts for the overall figure. This is important because many early tree surveys do not mention pollards (i.e., they are surveys only of *timber* trees), or do not detail their species. This said, when the species of pollards *are* specified they usually replicate in broad terms the balance of species of the timber trees. On John Darker's Northamptonshire estates in 1791, for example, ash made up 60 per cent of the total trees recorded, oak around 20 per cent and elm 20 per cent; when only timber is counted, the figures are 48 per cent, 25 per cent and 27 per cent respectively (NHRO YZ 2183). Similarly, on the Earle estate in north-east Norfolk in 1722, 66 per cent of the timber trees recorded were oak, 31 per cent ash and just under 3 per cent elm; the figures for the pollards were 70 per cent, 28 per cent and 1.2 per cent respectively (NRO BUL 11/283, 617X2).

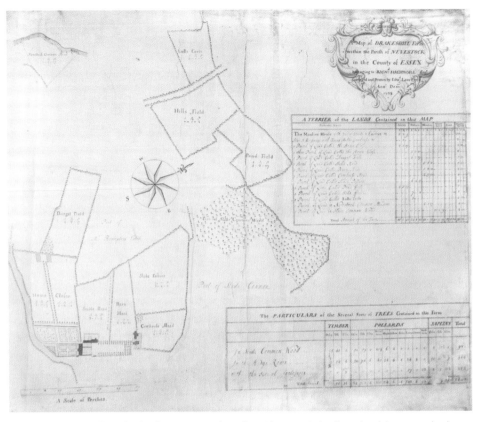

5.1 Map of Drakeshill Farm (evidently a minor gentleman's residence, to judge from the elaborate gardens), Navestock, Essex, surveyed in 1722. The panel bottom right lists the trees on the property, including the large numbers – mainly hornbeam pollards – growing on the adjacent area of Slade Common.

The overwhelming dominance of oak, elm and ash is remarkable given the fact that there are at least 25 indigenous or long-naturalised species capable of growing into reasonably sized trees, with a height of ten metres or more. The contrast with the balance of species that existed in the wild vegetation, before the advent of farming, is considerable. Although pollen evidence suggests that in many regions oak, elm and ash were important trees in the 'climax' vegetation, ash in particular was generally present at lower frequencies (and absent altogether from some areas) and elm was much more abundant, at least before the 'elm decline' of the early Neolithic (Rackham 2006, 82–90). Oak (accompanied by hazel) dominated the vegetation in the north, but large areas of birch–alder woodland also occurred on higher ground, and some areas were dominated by Scots pine (Gledhill 1994, 88–95; Jones 2012, 10–11). In most of England, certainly across the Midlands and the south, as we have seen, small-leafed lime (*Tilia cordata*) was the most important tree, accompanied by ash, oak, elm and hazel: species such as pine and yew were locally significant, together with alder on wetter ground (Bennett 1983; Bennett 1986; Peglar *et al.* 1989; Campbell and Robinson 2008). Pollen of pre-Neolithic date from Diss Mere in south

Norfolk, where hedgerow timber was later to be firmly dominated by oak, ash and elm, typically suggests that the 'wildwood' was characterised by *c*.40 per cent lime, 20 per cent oak, and 10 per cent each of alder, ash and elm (Peglar *et al.* 1989, 103).

Oak, elm and ash were not, however, dominant in the farmed landscape because they were in some way well suited to that environment. Most farmland trees were deliberately planted: even if self-seeded they needed to be consciously protected. In other words, which trees grew where, and in what numbers, was mainly a consequence of decisions made by individuals, or groups of people: farmers, landowners, land agents. The overwhelming dominance of these three species was thus largely a consequence of deliberate choice. Two main factors ensured their popularity: an ability to thrive in a wide variety of contexts and the wide range of uses to which their timber or wood could be put. Early writers always discussed these three trees before any others, sometimes followed by beech, walnut and chestnut. All agreed that oak made excellent structural timber: '*Oak* hath the preheminence of all others, for its strength and Durableness'; it was 'The best Timber in the World for building Houses, Shipping, and other Necessary Uses' (Meager 1697, 110). But it also makes good firewood, excellent charcoal, and cleaves easily, ensuring that it was particularly suitable for making floorboards and fencing, including park pales. It was also widely used by the joiner, while its bark was employed in tanning – an increasingly important consideration, as we have seen, in the eighteenth and nineteenth centuries. In Ellis' words, oak 'of all other Trees, claims the Priority of Regard in this Nation for its many transcendent uses' (Ellis 1741, 18). It was, in addition, catholic in its habits or, as Mortimer put it in 1707, it 'thrives best on the richest Clay, 'tho it will grow well on moist Gravel or the coldest Clay' (Mortimer 1707, 329). Blome thought that it would 'grow and prosper in any sort of Land, either good or bad' (Blome 1686, 250); Meager, that it would grow 'in any indifferent Land, good or bad, as Clay, Gravel, Sand, mixed, or unmixed Soils, dry, cold, warm or moist' (Meager 1697, 110).

Ash was different. While less sturdy, and so less useful as structural timber, it had many other uses. It was 'one of the most universal forms of Timber we have', according to Mortimer (Mortimer 1707, 336), while Timothy Nourse thought that it was 'a most useful wood to the Coach-maker, Wheeler, Cooper, and a Number of other Artificers', and that it had numerous uses on the farm, for fencing and bins, 'for Spittle and Spade Trees, for Drocks and Spindles for Ploughs, for Hoops, for Helves, and Staves, for all Tools of Husbandry, as being tough, smooth and light' (Nourse 1699, 119). Above all, its excellence as firewood was universally praised, for it burned well even when green. In Evelyn's words, 'the sweetest of our forest fuelling, and the fittest for ladies chambers' (Evelyn 1664, 40); while, to Moses Cook, 'Of all the wood that I know, there is none burns so well green, as the Ash' (Cook 1676, 76). Hartlib put it simply: 'Ash, for a hundred uses' (Hartlib 1651, 80). And on top of all this ash grew rapidly and, like oak, was not very choosy about *where* it grew. Although some early writers believed that it had some preferences, perhaps for 'light dry Mould', there was general agreement that it would grow on 'any sort of land', provided 'it be not too stiff, wet and boggy', although in general it seems to have been less prominent on more acidic soils (Mortimer 1707, 366).

Elm in its various forms also had a multiplicity of uses. Displaying good resistance to rot when waterlogged, it was thought 'proper for Water-works, Mills, Soles of Wheels, Pipes, Aquaducts, Ship Keels and Planks beneath the Water Line'. It was also used for making 'Axel trees, Kerbs Coppers … Chopping-Blocks … Dressers, and for Carvers work', as well as spades, shovels and harrows. Above all it made excellent boards and planks for floorboards, external weatherboarding and coffins (Nourse 1699, 115). Like oak and ash, moreover, elm could tolerate a wide range of conditions. While Ellis thought it preferred a 'damp or wettish soil', Mortimer believed it 'thrives best in rich black Mould', and did less well on sands and gravels, but he conceded that it would grow 'almost on any sort of Land' (Ellis 1741, 46; Mortimer 1707). Seventeenth- and eighteenth-century writers singled out another advantage: it caused 'the least offence to Corn, Pasture and Hedges of any Tree'. In part this was due to the fact that (unlike ash) its roots did not spread far, but it was also because it could be rigorously trimmed up as timber, so that it cast limited shade. Ellis thought that 'they don't damage any thing about them, as some other Trees do, whose Heads must not be trimmed up as these may' (Ellis 1741, 49).

As well as the almost universal dominance of the three species, also noticeable is the way that the relative balance between them displayed a degree of regional variation. We may begin by considering the old-enclosed 'woodland' areas lying to the north of the Thames, including much of East Anglia. Across this extensive tract of countryside oak (*Quercus robur* and *Q. petraea* are not distinguished in the documents) was generally the most common farmland tree. The extent of its dominance seems to have been greatest on the more acidic soils, as on the poor, often waterlogged lands of south Hertfordshire, south Essex and Middlesex. Here elm (*Ulmus procera*) often took second place, as on the manor of Boreham in south Essex in 1630 (oak 61 per cent, elm 29 per cent and ash 10 per cent) (ERO T/A 783/1) or at Bicknaire Farm, Great Baddow, in 1778 (oak 65 per cent, elm 23 per cent, ash 12 per cent) (ERO D/DRa C4). Of 1,782 timber trees recorded in Barnet in south Hertfordshire in 1786, 63 per cent were oak, 30 per cent elm and only 7 per cent ash (HALS DE/B983/E1), while of the trees valued at the enclosure of nearby Aldenham in 1803 most were in fact elm (51 per cent), with oak coming second (24 per cent), followed by ash (14 per cent) (HALS DE/X216/B2). Elsewhere in these districts, however, ash sometimes came second, as on an estate at Waltham Abbey in 1791, where oak made up 53 per cent of the trees, ash 25 per cent and elm 20 per cent, on a farm at Danbury and Purleigh in 1831, where no less than 82 per cent of the trees were oaks, with 15 per cent ash and only 3 per cent elm (ERO D/DOp E4), and on the Toppings Hall estate in Hatfield Peverel in 1791, where 48 per cent were oak, 36 per cent ash and 12 per cent elm (ERO D/DRa E23/6).

Oak was similarly dominant, and elm well represented, on the slightly acid loams found in parts of this broad region. In south-east Suffolk, on soils of the Burlingham 3 Association, oak made up 47 per cent of the trees auctioned from farms in the Aldham and Tattingstone area in 1810, elm constituting 28 per cent and ash 23 per cent (ERO D/DWe E3), while at Campsea Ashe elm was actually the most numerous tree recorded on a farm in 1807, making up 71 per cent of the total, although it accounted for only 6 per cent of the timber trees, of which there were few on the holding: oak made up 63 per cent of these (Ipswich RO HD11:475). On the dipslope of the Chiltern Hills, similarly, on the mildly acidic soils of the Hornbeam Association, oak accounted for around 65 per cent of the trees recorded in early surveys and never less than 45 per cent, here with elm and ash vying for second place. Where more fertile (but still acidic) loams were found, however, oak was unchallenged. On the Wick Association soils of north-east Norfolk, for example, of the 24,000 trees recorded on the Earle estates in 1722, 59 per cent were oak, with 37 per cent ash and a tiny 2.5 per cent elm (NRO BUL 11/283, 617X2); on the nearby Blickling estate in 1576, 63 per cent were oak, 36 per cent ash and less than 1 per cent elm (NRO NRS 8582 21 C2).

Oak was likewise the most numerous tree on the boulder clays of East Anglia, Essex and east Hertfordshire, especially where the soils were slightly acidic, on the more level areas of ground found in south Norfolk and north Suffolk. At West Bradenham in Norfolk in 1750 it made up 42 per cent of the 3,783 farmland trees valued, while elm and ash (undifferentiated in the survey) accounted for the remaining 58 per cent (NRO MS 9316, 7B9). Oak accounted for 70 per cent of the trees at Langley in Norfolk in 1676 (NRO NRS 11126,25E5) and made up over 80 per cent of the timber sold on a property in Whissonsett in the same county in 1762 (NRO HIL3/34/1–27, 879x3). But where the boulder-clay soils were more alkaline and better drained, further south in Suffolk, Essex and east Hertfordshire, oak was sometimes less prominent. In some places ash was the most important species, making up 69 per cent of the trees recorded at the appropriately named Badwell Ash in Suffolk in 1730, and 74 per cent of those at Ryes Farm in Little Henny in 1799 (Bury St Edmunds RO 613/642/1; ERO D/F 35/3/42). Elsewhere, elm (*Ulmus procera* or *carpinifolia*) took first place, as at Much Hadham in Hertfordshire in 1803 (69 per cent elm, 24 per cent ash and 7 per cent oak) or on Pearces Farm in Thorley and Sawbridgeworth in 1807 (71 per cent elm, 22 per cent oak, 8 per cent ash) (HALS Lob B/PC4/4; HALS DE/H/P11). At Sampford in Essex in 1843 no oak was recorded, but there were no timber trees, and the pollards (132 in all) were evenly balanced between elm (41 per cent) and ash (37 per cent), with maple making up the rest (ERO D/F 35/3/101). Elm was particularly prominent in west Suffolk, making up 55 per cent of the timber recorded on Thurston Hall Farm in Hawkedon in 1823–38, for example, and 63 per cent of that at Chevington in 1820 (Bury St Edmunds RO HA535/5/35; HA 507/2/460). In 1741 an anonymous traveller in this same district, between Bury St Edmunds and Stowmarket, described the hedges as 'full of timber trees, chiefly elm and some few oaks' (Wilson 2002, 265). But while on some properties on these more alkaline soils ash and elm might be the most numerous trees, oak was normally pre-eminent. At Finchingfield in Essex in 1805 it made up 90 per cent of the trees (ERO D/D Pg T8); at Little Walden in the same county in

1826, 71 per cent (ERO D/F 35/3/42); at Olive's Farm in Hunsdon, Hertfordshire, in 1556 92 per cent (PRO/TNA E315/391); and on a farm in Colliers End in the same county in 1794, 93 per cent (HALS DE/B1768 P2).

In the old-enclosed areas south of the Thames oak was similarly almost everywhere the dominant tree. In Sussex, according to Young, the soil was 'very congenial' for its growth: 'It is a weed which springs up in every protected spot' (Young 1808, 169). The Weald was 'an oak district', thought Stevenson, but 'this tree is to be found in almost every other part' of Sussex (Stevenson 1809, 441–2). The hedges of a field on the Leeds Abbey estate in Kent contained in 1718 there were 33 oaks but only 13 ashes and 3 elms (KHL C82/1–69), and oak made up 95 per cent of the farmland timber growing on the Sheffield Park estate on the Kent and Sussex border in the late seventeenth century (KHL U269 E328/2). The timber valued in 1809 on farms owned by the Earl of Camden in Kent comprised 4,190 oak but only 133 other trees (KHL U840 EB 116); some of the trees were probably growing in woodland, but the pattern is clear enough. Across the whole of the old-enclosed region of the south-east and East Anglia oak was thus usually the most important farmland tree, or at least came a close second to ash or elm. Only in restricted localities was it regularly a minor feature of the landscape. In some coastal areas, in particular, elm – which has good resistance to salt – could be overwhelmingly dominant. In 1753, on an estate on the Isle of Thanet in Kent, elm made up 99 per cent of the 4,119 trees recorded (KHL R/U438/E35). But on inland sites, and especially on the more acid soils, oak was king.

In the old-enclosed districts in the west of England the situation was similar. In Cheshire the 'greatest part' of the hedgerow timber was oak (Holland 1808, 197). In Herefordshire there were many 'fine trees of oak and elm' – 'even in the red and comparatively barren soils the oak flourishes with astonishing luxuriance' (Duncumb 1805, 95); on the Clehonger estate in 1820, for example, 72 per cent of the farmland trees recorded were oak, 16 per cent were elm and 9 per cent ash (HAR C38/49/3/1). In Shropshire the hedgerow timber consisted of 'oak and ash principally' (Plymley 1803, 212); oak made up 75 per cent of the mature trees growing on the farmland on the Oakley Park estate in 1806, ash 16 per cent and elm 4 per cent; the timber felled on a farm at Bausley, Alberbury, in 1795 comprised 70 per cent oak, 23 per cent ash and 4 per cent elm; while oak constituted two-thirds of the timber sold at an auction of farmland trees at Chirbury in 1843 and 67 per cent of the 1,921 trees auctioned on lands around Shifnall in 1798 (with ash 21 per cent and elm 6 per cent) (SA 552/18/4/61; 665/3/479; 631/2/138; 5735/3/3/15). As in other counties, there were many local variations in, and deviations from, the overall pattern, especially on small farms. At Little Cowarne in Herefordshire in 1811, for example, 84 per cent of the trees on a property were in fact ash, with only 15 per cent oak and 1 per cent elm (HAR C38/49/3/1). But oak usually predominated in Shropshire and Herefordshire, while in Warwickshire the old-enclosed areas of the north, centre and west of the county contained a 'great quantity of the most valuable oak', according to Murray in 1813. But south-east Warwickshire was part of the 'champion', much of it still open or only recently enclosed when Murray was writing, and here in contrast 'Elm may be considered the weed of the county' (Murray 1813, 140–41). In the adjacent county of Gloucestershire, similarly, oaks grew in 'several parts' of the Vale of Evesham, but 'the elm-tree grows in almost every district' (Rudge 1807, 241).

This was part of a much wider and slightly mysterious pattern, for in most parts of the 'champion', at least in the south and Midlands, oak was routinely a far less important tree than in old-enclosed districts, frequently coming second or even third to elm and ash. Oak thus accounted for only 19 per cent of the trees surveyed at Wilby in Northamptonshire in 1764, 15 per cent of those recorded when Finedon was enclosed in 1806, and less than 1 per cent of those noted at the enclosure of Irthlingborough in 1808 (NHRO B(G)1; NHRO ZA898; NHRO ZA 906). A survey of the manor of Milcombe in Oxfordshire, made in 1656, lists almost no oak but many ash and elm growing on the premises of the tenants (NHRO C(A) Box104 4 1656). It is noteworthy that when sixteenth- or seventeenth-century leases from Midland districts stipulated that the tenants should plant trees on their holdings, elm and ash are more frequently mentioned than oak. One for a property in Slapton in Northamptonshire, for example, drawn up in 1651, stipulated that the lessor should plant 'yearly four ash or elm' on his holding (NRO ZB 1883/06); another, for Hellidon in 1595, that one 'good or sufficient plant of ash' should be planted each year until the messuage was 'completely stored' (NHRO HOLT 638).

Even after the open fields had been enclosed and the land lay in hedged fields, oak was still often poorly represented. When the Duke of Powis' extensive estates around Heyford, Glassthorpe and Newbold in Northamptonshire were surveyed in 1758, *only* elm and ash were specifically noted (NHRO ZB 1837), while oak was the third most common farmland tree in a quarter of the 16 Northamptonshire parishes in which the Montagu estate held property in 1749 and the second in all but two of the others (Boughton House archives). In some of these places elm was the most numerous species, but usually ash was in a clear majority, as it was on John Darker's properties in the same county in 1791, where 59 per cent of the trees recorded were ash, 21 per cent oak and 20 per cent elm (NHRO YZ 2183). This secondary (or tertiary) importance of oak as a farmland tree in Northamptonshire does not appear to have changed significantly in the late eighteenth or even the early nineteenth century. It accounted for only 20 per cent of the trees growing on properties at Staverton in 1835, 6 per cent at Welton in 1839 and around 8 per cent at Teeton in 1812 (NHRO ZB 887; NHRO ASL 392; NHRO BH(H) 461). In the former 'champion' parts of west Norfolk, similarly, on the edge of the Fenland around Wimbotsham and Downham Market, oak made up only 36 per cent of the trees on the Stradsett estate in 1813 and 1814, with elm contributing 31 and ash 33 per cent, while lists of timber felled on the same property in 1813 and 1820 record slightly more elm than ash, and no oak at all (NRO BL/BG 5/3/4). On the estate properties in Marsham, Fincham and Shouldham 53 per cent of the 1,062 trees were ash, 24 per cent were elm and only 23 per cent oak (NRO BL/BG 5/3/4).

Yet, to complicate matters further, such records as we have for the north of England suggest that the dominant farmland tree here was oak, even in 'champion' or former

5.2 View from the Cleveland Hills, Yorkshire, showing a landscape of hedges studded with oak trees. Oak has long dominated the landscape in many northern areas, where elm was always rare.

champion areas, and that elm was rare (Figure 5.2). On an estate at Ellerton in the Vale of York in 1768 there were 1,015 oaks but only six ash and no elms (Bev DDX 73/51); at nearby Thorne in 1818 oak made up 72 per cent of the trees, ash 21 per cent and elm 5 per cent (Bar EM 291); while at Temple Newsam near Leeds in 1825 the figures were 50 per cent, 38 per cent and 2 per cent respectively (Lds WYL189/B/61). Overall, oak constituted around 72 per cent of farmland trees recorded in seventeenth-, eighteenth- and nineteenth-century surveys in Yorkshire as a whole, ash 25 per cent and elm a mere 2 per cent.

These regional variations were evidently the outcome of the complex interplay of a range of factors. Some were evidently natural in character. English elm (*Ulmus procera*) is on the edge of its natural range anywhere north of Derbyshire and while wych elm (*Ulmus glabra*) will tolerate the northern climate it requires deep, rich and fairly alkaline soils to flourish and is slower growing, while its timber is harder to work. But the explanation for other variations is less clear, especially the secondary importance of oak in many Midland 'champion' districts. This is especially striking given that the *woodland* in these same areas was dominated by oak. We noted in Chapter 3 how in such landscapes the majority of trees (other than willows) were, prior to enclosure, generally concentrated in the tofts and crofts of the nucleated villages. It may well have been that oak, which develops a wide, dense crown at an early stage of growth, cast too much shade in such locations. Ash had a more open crown and, although many early writers commented on how it robbed nutrients from the adjacent fields, spreading its roots widely, such disadvantages were presumably outweighed by its numerous uses, especially as fuel. Elm made a particularly good choice where trees needed to be tightly packed along the hedges surrounding small village closes both because its roots did not spread far and also because it could be rigorously trimmed up as a timber tree, so that it cast limited shade. Ellis commented that he 'never saw so many grow so large and flourish in so little room as these will, even almost close together' (Ellis 1741, 46). While all this may be true, it is strange that, while oak certainly increased in importance in seventeenth- and eighteenth-century enclosure hedges, it seldom attained the levels seen in many old-enclosed regions. Perhaps long-established planting traditions were difficult to change. Like so many questions about early tree populations, this one remains intractable and requires further research.

The 'minority' trees of farmland

Other kinds of tree are recorded by early surveys growing in small numbers in fields and hedgerows alongside oak, ash and elm. These include maple, hornbeam, aspen (*Populus tremula*), black poplar (*Populus nigra* var. *betulifolia*), lime, alder, sycamore (*Acer pseudoplatanus*), beech, holly, sweet chestnut (*Castanea sativa*), walnut (*Juglans* spp.), wild service and various kinds of willow, together with fruit trees such as apple and cherry. Indeed, only rowan, whitebeam (*Sorbus aria*) and birch were regularly shunned by hedgerow planters, appearing at very low levels only in some western and northern districts. In general, these 'minority' species made up less, and often much less, than 10 per cent of the total number of trees present on a property. In the county of Essex, for example, they constituted 4 per cent of those surveyed on the Toppings Hall estate, Hatfield Peverel, in 1791 (ERO D/DRa E23/6), and less than 1 per cent of the 1,334 timber trees recorded on Sir John Henniker's estate in Waltham Abbey in 1791 (Ipswich RO HA 116/5/11/2). They accounted for only one of the 439 growing on a farm at Little Walden in 1826, while at Bicknaire Farm in Little Baddow in 1778, at Finchingfield in 1772, at Ryes Farm in Little Henny in 1799 and on a farm at Danbury and Purleigh in 1831 no examples at all were recorded (ERO D/DRa C4; ERO T/A 783; ERO D/F 35/3/42; D/DOp E4). There were, it is true, some exceptions to this broad rule. The 'minority' trees accounted for 14 per cent of those at Curd Hall, Coggeshall in 1734 (ERO D/Dc E15/2) and a surprising 61 per cent at Navestock in the south of the county in 1722 (ERO ERO D/DU 583/2). Such exceptions, here and in other counties, generally come from slightly earlier documents, and it is possible that there was an increasing emphasis on oak, ash and elm over time. It is noteworthy that the timber accounts of Richard DuCane, a landowner around Coggeshall in the 1730s, show him cutting down ash, oak, elm, maple and hornbeam, but never replanting the latter two species (ERO D/Dc E15/2). Nevertheless, there seems little doubt that from the Middle Ages oak, ash and elm had always been the dominant trees on the vast majority of holdings, not only in Essex but throughout the country.

As already intimated, 'minority' trees were more common as pollards than as timber, and for this reason they may be under-represented in our sources, which often fail to

5.3 An old pollarded hornbeam in a hedge in south Norfolk. Old hornbeam pollards are more common in the landscape of East Anglia than we might expect from early tree surveys from the region.

specify the species of pollarded trees. It is noteworthy that seventeenth- and eighteenth-century writers such as Thomas Hale discuss only a limited range of timber trees, mainly oak, ash and elm, but suggest that 'there is scarce any Tree that that may not be brought to a Pollard at the owner's Pleasure' (Hale 1756, 321). Yet, as we have also noted, for the most part, when surveys do specify the species of pollards, these broadly replicate the balance of species of the timber trees, and are likewise dominated by oak, ash and elm. The relative rarity of these other species is thus evidently real and indicates that they had fewer uses than oak, ash or elm, grew more slowly or were more demanding in their requirements. In some cases, however, it reflects the fact that contemporaries thought that they were better coppiced, in woods or hedges, than grown as hedgerow trees.

Some of these trees had restricted, patchy distributions. Lime, presumably in most cases small-leafed lime, was recorded sporadically in old-enclosed districts, especially in the east, and most frequently in Suffolk and Essex. Lime made up nearly a third of the trees recorded at Navestock in 1722, but this was an extreme case: at Chevington in Suffolk 1820 it constituted 5 per cent of the total, as it did at Frostenden and Uggeshall in the same county in 1849, while it made up only 2 per cent of trees at Toppings Hall, Hatfield Peverel, Essex in 1791 and a mere 0.2 per cent of the trees recorded at Cretingham and Brandeston in Suffolk in 1821 (ERO D/DRa E23/6). Elsewhere it was generally present in only tiny quantities. There were only three limes listed among the 4,323 trees on the properties of the Earl of Camden in Kent and Sussex in 1809 (KHL U840 EB 116), and only three were recorded among the 24,000 trees described on the Earle estates in north-east Norfolk in 1722 (NRO BUL 11/283, 617X2). Hornbeam was also a tree of old-enclosed, 'woodland' areas, although almost exclusively in the south-east of the country and East Anglia. It was almost invariably pollarded: its wood was so hard and difficult to work that it was of little use as timber. Indeed, as late as 1838 Loudon suggested that the only hornbeam trees of any size were pollards or former pollards (Loudon 1838, 2005). On rare occasions hornbeam could make up as much as 15 per cent of the trees growing on farms, as at Navestock in Essex in 1722 (ERO D/DU 583/2), but it was usually present in small amounts. It thus accounted for only 0.3 per cent of the 3,630 trees growing at Curd Hall, Coggeshall, Essex in 1734, where it was described by its alternative name of 'Hard Beech' and, unusually, was mainly present as timber (ERO D/Dc E15/2). It constituted 1.2 per cent of the trees listed on a farm at Finchingfield in Essex in 1805 (ERO D/D Pg T8) and around 1 per cent of the old pollards sold at Tilgate Farm, Broxted, in the same county, in 1843 (ERO D/F 35/3/135). It might have been slightly more common in the seventeenth- and eighteenth-century countryside than the documentary sources suggest, for it is not that rare today as a former pollard in the hedges of south Norfolk, for example, a district in which it is seldom referred to in documents (Figure 5.3). As we shall see, hornbeam was much more important as a tree of wood-pastures than it was of enclosed farmland.

In contrast to these trees, almost always present only at low densities, was maple. This was moderately common in many old-enclosed districts, often making up between 1 and

5.4 A rare example of a maple pollard, typically cut low, on a short bole. Maple was traditionally rare as a hedgerow tree, and was more commonly treated as a pollard than a standard.

4 per cent of the total number of trees, but was sometimes recorded in greater numbers. It accounted for 12 per cent of trees at Curd Hall, Coggeshall in Essex in 1734; 22 per cent of those sold at Little Sampford in the same county in 1843; and 21 per cent of those surveyed at Lindsey in Suffolk in 1733 (ERO D/Dc E15/2; ERO DIF 35/3/101; Ipswich RO B E3/5/1). In contrast, it was relatively rare in the champion Midlands, accounting for less than 0.1 per cent of trees recorded in the majority of surveys and never reaching more than 4 per cent. It was sporadic also in the north of the country. It is often described as a tree of calcareous soil, and a number of fine 'veteran' examples survive – mostly growing on former common land – in the Cotswolds. But maple turns up in some numbers in other locations. Drakeshill Farm, in Navestock in Essex, occupies acidic clay soils but the trees growing in the hedges and fields there in 1722 included 34 examples of maple. Maple sometimes appears as a timber tree, and it had a range of specialised uses, for making

dishes, spoons and 'other curious Turner's Ware', as well as musical instruments, which required pieces of wood larger than would be provided by poles cut from pollards. But it is more usually recorded as a pollard, as at Aldenham in Hertfordshire in 1803 or Kelshall in the same county in 1727, where 53 specimens were among the many trees growing on a property of less than 40 acres (16 hectares), 36 of them in one hedge of a small (4-acre) field (HALS DE/X 216 B2–4; HALS DE/Ha/B2112). Surviving examples suggest that it was usually lopped at a low height, sometimes less than two metres (Figure 5.4). Ellis explained that its wood was brittle, light and soft, and that it was '[n]ot so profitable to burn as some are. They are sometimes made Pollards, but make a slow Return that Way: in standards they seem to do better, because they are not subject to those Evils that the Pollard is: for this being a soft Wood, is apt to let in the Wet after topping' (Ellis 1741, 84).

Maple was presumably common in the past, as it is today, as a shrub in hedges, where it seeds relatively easily. It is also – usually in combination with hazel and/or ash – a frequent component of the coppiced understorey in woods, especially in the Midlands and south. Farmers and landowners evidently preferred to manage it as underwood rather than as a pollard: if it self-seeded in a hedge it was usually plashed or laid there with the rest of the shrubs. According to Moses Cook,

> If you let it grow into Trees, it destroys the wood under it; for it leaves a clammy Honey-dew on its Leaves, which when it is washed off by Rains, and falls upon the Buds of those Trees under it, its Clamminess keeps those Buds from opening, and so by degrees it kills all the Wood under it; therefore suffer not high Trees or Pollards to grow in your Hedges, but fell them close to the Ground, and so it will thicken your Hedge, and not Spoil its Neighbours so much. (Cook 1676, 99)

The trees so far described were all indigenous, or had at least been present in England since prehistoric times, but some of those planted in seventeenth- and eighteenth-century hedges were introductions from abroad. Walnut (*Juglans regia*) had probably been brought to England in Roman times, to be joined by the black walnut (*J. nigra*), an American tree, in the seventeenth century: documents do not distinguish the two. Neither propagates itself easily in the wild: they are extremely demanding and are intolerant of competition (Savill 2002, 46). Walnut was highly valued by elite writers, especially as a furniture wood, before large-scale imports of mahogany from the Americas began in the early eighteenth century. It was also widely used to make gunstocks. But it did not produce very good firewood and was thought by many to harm the hedges in which it grew. Ellis also warned that 'boys and others' would damage hedges while trying to steal the fruit (Ellis 1741, 63). Nevertheless, it is regularly recorded as a farmland tree, if at low frequencies, and, unlike the trees so far discussed, was as common in 'champion' and former champion areas as it was in old-enclosed districts. On John Darker's estates in Northamptonshire in 1791 walnut was the fifth most common species recorded, after ash, oak, elm and poplar, although in total accounting for only 0.3 per cent of the trees on the estate (NHRO ZB 1837). On the Boughton estate in the same county it was present at similar levels in 1749 (Boughton House archives). But it was occasionally more prominent in the county, as on a farm at Teeton in 1812, where it made up nearly 6 per cent of the trees (Boughton House archives; NHRO B(H) 461). In Gloucestershire, Rudge described in 1807 how local stocks had been depleted by the demands of the Birmingham gunsmith industry:

> Only here and there is a solitary walnut-tree seen growing. In the parish of Arlingham
> there are more, perhaps, than in many other parishes combined. So abundant, indeed, is
> the fruit this year (1805) that it has become an article of commerce, and two vessels are now
> (October 11) being laden with walnuts for Scotland. (Rudge 1807, 245)

Walnut was also found in old-enclosed areas, both to the east and the west of the 'champion'. It made up 44 of the 24,000 trees on the Earle estates in north-east Norfolk in 1722 (NRO BUL 11/283, 617X2) and 0.8 per cent of those on a property at Finchingfield in Essex in 1805 (ERO D/D Pg T8). But it could be more numerous where particular owners planted it enthusiastically: it formed 5 per cent of the trees at Navestock in Essex in 1722, for example (ERO D/DU 583/2). At Curd Hall, Coggeshall, in the same county, it accounted for only 0.8 per cent of the trees recorded in a survey of 1734, but its numbers increased rapidly during subsequent years, the owner Richard Du Cane recording in his forestry notes how he planted 120 examples on this and two other farms he owned in the locality in 1734; a further 112 in 1735; and another 50 in 1736. Almost all, it should be noted, were established in the hedges of the farm fields, rather than in the gardens or orchards (ERO D/Dc E15/2). Du Cane was not alone. In 1815 one Kentish landowner described how he had planted 100 walnut trees in the hedges on his farm (KHL U3750/21/1).

Sycamore was also, almost certainly, an introduced species, although unlike walnut it is naturalised and once established in a locality will seed freely. Indeed, some have argued that it is, in parts of northern and western Britain, a native (Green 2005): in 1776 Pennant reported scattered groves of sycamores in the Lake District. It is certainly true that, in the Pennines especially, some ancient examples, usually former pollards, can be found (Figure 5.5). But it is surprising that in Yorkshire, where it is often described in documents as 'plane', sycamore never seems to exceed 1 per cent of the total numbers of trees recorded in valuations, surveys and sales documents before the nineteenth century. One of the earliest records is from Healaugh in 1751, where eight specimens were recorded on the farm of John Adcock (Lds WYL68/33). Even in the five years of timber accounts for Newby Hall spanning the years 1829–34 it is mentioned only three times (Lds WYL5013/2234). Further south, in 'champion' Northamptonshire, it is recorded on the Montagu estate from 1749, but again only at low levels (less than 2 per cent) (Boughton House archives). No examples at all are mentioned in a survey of over 3,000 trees on the Duke of Powis' Northamptonshire estates around Heyford and Newbold in 1758, nor among the 4,500 trees recorded on John

5.5 A magnificent sycamore pollard at Eyam, in the south Pennines. Old sycamores like this are rare in the south of England, but relatively common in many northern and upland areas.

Darker's estates in the same county in 1791 (NHRO ZB 1837; NHRO YZ 2183). Of the 983 trees recorded in the valuations of timber associated with the enclosure of Wilby in 1806, a mere four were sycamores.

In old-enclosed areas, both to the west and to the south-east of the champion, it was no more common. In Shropshire, for example, it accounted for only eight of the 1,921 trees auctioned on farmland around Shifnall in 1798; a mere 1.4 per cent of 6,160 mature farmland trees on the Oakley estate in 1806; and 2.6 per cent of trees sold from the fields and hedges at Kilsall in 1836. On the Clehonger estate in Herefordshire in 1820 there was a single sycamore out of 1,386 trees (HAR C38/49/3/1). One example appears among the 53 trees valued on a farm on the Leeds Abbey estate in Kent in 1718 (KHL C82/1–69), but only three were listed among the 4,323 trees on the properties of the Earl of Camden in Kent and Sussex in 1809 (KHL U840 EB 116). In Essex it made up 2 per cent (1 out of 50) of the trees at Toppings Hall in 1791, while at Little Walden in 1821 a single example was recorded out of a total of 671 trees, and only one (a pollard) was included among the 378 trees sold at Tingate Farm, Broxted, in 1852 (ERO D/F 35/3/135).

In parts of East Anglia it may have been slightly more common. Five out of the 352 trees recorded on a farm at Bardwell in Suffolk in 1730 were sycamore, and at Chevington in the same county there were two (one a pollard, one timber) among 264 trees listed in 1820 (Bury St Edmunds RO B E3/10/10.5; Bury St Edmunds RO HA 507/2/460). In Norfolk there is a reference, as early as 1613, to the planting of 'young sikamores' beside a ditch at Thetford (MacCulloch 2007, 99). Nevertheless, it generally appears in surveys before the nineteenth century as a tree of parks and gardens (as at Hunstanton in north-west Norfolk in the 1770s: Williamson 1998, 253). Not a single specimen was recorded among the 24,000 hedgerow trees on the Earle estates in 1722 (NRO BUL 11/283, 617X2). Overall, sycamore thus appears to have been uncommon as a tree of the countryside and many seventeenth- and eighteenth-century writers on farming and forestry either fail to mention it at all or regard it as an ornamental, found in gardens and parks. While some nineteenth-century commentators did recommend that it should be planted in hedges this was only the case 'if ornament be studied' (Nicol 1799, 249). Ellis and Langley both describe how its wood and timber could be used in similar ways to those of maple, but noted that it was of the 'soft, woody tribe', and made only moderate fuel (Ellis 1741, 87; Langley 1728, 153–5). There was probably little point in planting it in preference, in particular, to ash, to which several writers compared it, finding it inferior, if by no great margin.

Poplar and aspen are often noted in early surveys: in almost all cases references to the former seem to be to the black poplar, rather than the white poplar (*Populus alba*), or 'abele'. Both trees grow with remarkable speed, and their wood is light and easily worked but tough, and excellent for making boards for flooring or for the exterior of buildings, and for joists and chair frames (Cooper 2006, 24–8). Aspen was used to make wagon bottoms, arrow shafts, oars and clogs. Its charcoal was thought to be good for manufacturing gun powder, while its bark could be used for tanning. Both woods, moreover, have mildly fire-resistant qualities, so that they were often used in mills and other industrial premises. Indeed, a survey of surviving and historically documented examples of black poplar in Norfolk suggests that over a third stand beside the sites of mills, smithies, kilns and malthouses, where they were presumably planted to provide material for repairing floors, doors and the like (Barnes *et al.* 2009). Neither wood thus makes good fuel although both – and especially black poplar – were often managed as pollards. Most examples of black poplar were deliberately planted. Propagation was generally by cuttings, as male trees vastly outnumber female ones, ensuring that seeds are in short supply (Cooper 2006, 20–22). Early writers classified both trees as 'aquaticks' but only the black poplar was (and is) really a tree of damp ground. Aspen, in fact, was most numerous on slightly acid loams and clays, such as the clay-with-flints of the Chilterns. Ellis believed that aspen did best in 'low, moist places', but noted that it also grew reasonably well in dry locations, in the hedges of his own farm and 'on our high Common' in Gaddesden, both in the Chilterns (Ellis 1741, 96). It is frequently mentioned in documents relating to Shropshire, always in places with freely draining, slightly acid loams. A lease from Ludford from 1683 ordered the tenant not to pollard any 'oak, ash, aspen or crab tree fit to make timber' (SA 11/535), and it is recorded in a number of sales and valuations, although never in large quantities. At Oakley in 1806, for example, aspen made up 0.4 per cent of the 6,160 trees listed on the estate farmland (SAR 552/18/4/6). It appears only at low frequencies on other soils, such as the East Anglian boulder clays: at Monewden in Suffolk in 1821 it accounted for only 0.1 per cent of the trees described (Rackham 1986, 218) and on the Henneker estate at Waltham Abbey there were two aspen trees out of the 2,812 whose species was specified, outside the woods and copses (Ipswich RO HA 116/5/11/2).

The black poplar, in contrast, was mainly a tree of damp soils, especially on alluvial flood plains. Notable concentrations still exist on low-lying wetlands in Cheshire and Shropshire. In the seventeenth and eighteenth centuries it also grew, to a greater extent than today, on clay soils, especially calcareous ones, in places where the drainage was poor (Cooper 2006, 9–12). On the Broxbourne Mill estate in east Hertfordshire in the late eighteenth century poplars made up 43 per cent of the trees on the 'enclosed meadows' beside the river Lea, but also 4 per cent of the trees recorded in the 'inclosure land' on the boulder clays (HALS B479). It could be found regularly if in small quantities across the claylands of Suffolk and Essex, often apparently as a hedgerow tree, making up 0.6 per cent of the trees recorded at Bardwell in Suffolk 1730 (Bury St Edmunds RO B E3/10/10.5), 1.1 per cent of those at Campsea Ashe in 1807 (Ipswich RO HD11:475), 2 per cent of those mapped at Lindsey in Suffolk in 1733 (Ipswich RO B E3/5/1) and an uncertain but perhaps significant quantity at Polstead in Essex in 1407 (Rackham 1986, 218). There were also 'a few' on a farm on the Flixton estate in north Suffolk in 1874, far from any flood plain (Lowestoft RO HA12/D3/8/14). Black poplars were, for the most part, thinly scattered across 'champion' districts – there were 26 among the 982 mature trees listed at the enclosure of Finedon in Northamptonshire in 1806, and two were recorded (out of 888 trees) at Staverton in the same county in 1835

5.6 A pollarded black poplar growing in a hedge at Quarrendon, just north of Aylesbury. Black poplars are a particular features of the landscape of the Vale of Aylesbury.

(NHRO ZA898; NHRO ZB 887). But in the Vale of Aylesbury in Buckinghamshire, and extending into the neighbouring areas of north Hertfordshire, a remarkable collection still survives, mainly on damp clay or alluvial soils: indeed, around half of the surviving known population of the species in England is to be found here (Cooper 2006, 44). There is some association with areas of early enclosure – particularly fine old pollards exist in the hedges around the deserted village of Quarrendon, just to the north of Aylesbury – but many specimens appear to grow in relatively recent, eighteenth-century hedges (Figure 5.6).

Quite why so many poplars were planted in this district remains unclear. Evidently a tradition arose, and was perpetuated, of using this species in preference to the willows that were more typical of 'champion' countryside in the Midlands, not only on the wide flood plains of the region but also more widely in the open fields or in enclosures made from them. Sixty-two per cent of the 3,055 trees recorded in the claims submitted when Irthlingborough in Northamptonshire was enclosed in 1808 were willows (the sources do not distinguish clearly between crack willow [*Salix fragilis*] and white [*Salix alba*]). No fewer than 410 of the 1,038 recorded in claims made at the enclosure of Finedon in 1806 were so described, as were a third of the 762 valued at Wilby in 1764 (NHRO ZA898; NHRO X1657; NHRO ZA 906). Ellis described how willows were common 'either in standards, pollards or in hedges' in the open-field districts of north Hertfordshire, around Baldock, and commented that the loppings were 'of such great use, that I have known it the only Wood they have in some parts of Rutland in their Open Fields' (Ellis 1741, 106).

The poles from pollards were used, into the twentieth century, to make hurdles and rough fencing, large amounts of which were required to fold sheep and otherwise control stock in the largely hedgeless landscape of the open fields. Willow poles also, according to Mortimer, made good stakes and tool handles, as well as faggots for burning, while larger pieces were used by turners (Mortimer 1707, 364). Most willows were cut on a short rotation of three to four years, 'just before Winter, or in the Spring' (Ellis 1741, 106).

Although a particular feature of the champion Midlands, willow also, of course, grew widely on meadows and flood plains in old-enclosed districts, although here alder was perhaps more common, especially in the east of England, as it was across large parts of the north. In general, alder did better where wetland soils were peaty, rather than alluvial. On a farm in Scarning in Norfolk in 1764 it constituted less than 1 per cent of all trees, but in 1813 at Brick Kiln Farm at Reepham, a short distance away, much of which lay on a flood plain, it made up no less than 38 per cent of the total population (NRO BCH 20; NRO MC 687/29–31 813 x 3). Alder was mainly, however, grown in woods or 'carrs' on damp ground, or as a component of the coppice within wet areas in woods on heavy clays. It was much valued for its durability in waterlogged conditions, and was thus widely used for jetties and river-bank revetments. It was also used for scaffolding poles – when Blickling Hall in Norfolk was constructed in the early seventeenth century payments for alder poles are among the first to appear in the building accounts (NRO MC3/47) – but was perhaps most important as a source of charcoal. In addition, its bark could be used for tanning and its wood was in demand for a range of specialised uses, including clog making.

The large numbers of willows growing in parts of the 'champion' Midlands represent one of the main exceptions to the general dominance of the farming landscape by oak, ash and elm. But in certain areas this dominance was also challenged by something that has since virtually disappeared from the countryside. At low frequencies, apples, cherries and pears often appear as farmland trees in early surveys, from Yorkshire southwards, although rarely in areas of heavy clay: they were instead characteristic of slightly acid loams. But on such soils they were in certain districts more widely planted. Norden in 1608 commented on the abundance of fruit trees in the hedges of Devon, Gloucestershire, Kent, Shropshire, Somerset and Worcestershire, as well as in many parts of Wales. Beale thought that Herefordshire hedges 'in most places' were planted with apples (Beale 1657, 2), Nourse in 1700 reported that fruit trees were widely planted in hedges in 'Herefordshire, Worcestershire, some parts of Gloucestershire' (Nourse 1699, 28), while Plot recorded the same practice in parts of Staffordshire in 1686 (Plot 1686, 384). In Shropshire apples appear in a number of surveys of farmland trees from the south of the county – 72 'crab apple, apple trees and beech' were marked for felling at Roughton near Bridgenorth in 1720 (SAR 2028/1/5/31) – and leases frequently refer to them. One for land at Whitton and Ludford from 1653 instructed the tenant to plant 'ten oaks, ten ashes and ten apple trees' each year (SA 11/96); similar injunctions appear in leases from Whitton in 1660 (SA 11/456), Woodbatch in 1661 and 1724 (SA 11/565/6) and Bishop's Castle in 1732 (SA 11/560–61).

Others, such as examples relating to farms at Ludford in 1683 or Bishop's Castle in 1728, ordered that 'No oak, ash, aspen or crab trees fit to make timber to be cut at the butt' (SA 11/535; 11/537: the crabs were presumably used as rootstock for grafting). At Bishop's Castle in 1654, similarly, the tenants could take wood for hedgebote and firebote, 'crab trees excepted' (SA 11/497).

In some western districts fruit trees were still being maintained in hedges in large numbers up to and beyond the end of the eighteenth century. In his *General View of the Agriculture of the County of Hereford* of 1794 Clarke referred to debates in the county over the relative advantages of planting apple trees in orchards or 'in the arable fields' (Clarke 1794, 36); in 1807 Rudge described how 'apple and pear trees are often seen growing in the hedge-rows' in Gloucestershire, a practice he criticised as a 'temptation to theft and plunder' (Rudge 1807, 197, 213); while in Worcestershire Pitt noted in 1810 how 'the various kinds of fruit-trees are often dispersed over the country in hedge-rows' (Pitt 1810, 148). But the impression is that, by this stage, fruit production in these areas was becoming concentrated exclusively in orchards. Other counties had already made this transition. There are thus no references to fruit trees in the hedges of Kent by the end of the eighteenth century, while Norden in 1608 was already lamenting the fact that they were gradually disappearing from the hedges of Middlesex and Hertfordshire, as the modern generation failed to replace those that had grown old and died (Norden 1608, 201). In the case of Hertfordshire, where the planting of fruit trees in hedges appears to have been restricted to the Chiltern dipslope

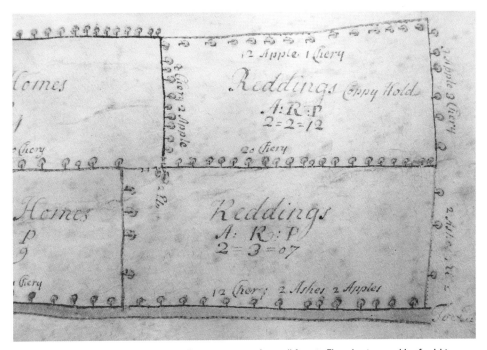

5.7 Detail from an undated late seventeenth-century map of a small farm in Flaunden in west Hertfordshire, showing hedges crammed with fruit trees.

in the west of the county (and extending into the adjacent parts of Buckinghamshire), this decline was, in fact, gradual. Moses Cook, a Hertfordshire man, was still advocating the planting of fruit trees in hedges in 1676 (Cook 1676, 138) and an undated late seventeenth-century map of a *c.*10-hectare farm in Flaunden near the Hertfordshire/Buckinghamshire border details 27 oaks, 18 elms and 9 ash growing in the hedges, and 15 'asps' (aspen), all outnumbered by 59 apple trees and no fewer than 165 cherries (Figure 5.7) (HALS DE/X905/P1). William Ellis' writings in the 1740s suggest that cherry trees in hedges were a normal feature of the landscape, both in his native west Hertfordshire and across the county boundary in Buckinghamshire. In 1749, however, while Pehr Kalm observed how fruit trees could be seen 'in hedges around the inclosures' in west Hertfordshire, he implies that they were uncommon (Lucas 1892, 147). The planting of fruit in hedges was probably fading out by this time. In neither Hertfordshire nor Middlesex is the practice referred to after the late eighteenth century.

To an extent, the prominence of fruit trees in particular districts was a function of natural factors, especially the presence of slightly acid loamy and reasonably freely draining soils. Moses Cook noted in 1676 how wild cherry was common in the woods of west Hertfordshire and seeded spontaneously in the hedges there. But in the last analysis it was the decisions of local people that ensured that hedges were filled with fruit trees rather than with oak, ash or elm. Individual landowners, knowledgeable about the local environment, maximised the profits they could make from it, profits which might derive from the wood and timber, as well as from the fruit, of these trees. Ellis considered that cherry timber was comparable in durability and strength to that of oak, although modern foresters note its variable quality in this regard (Ellis 1741, 65–6). Documentary sources confirm this dual importance, valuations of trees made at the enclosure of Aldenham in south Hertfordshire, for example, distinguishing between 'cherry' and 'cherry timber' (HALS D/EX 216 B2–4). Apple wood burns well, although it was generally agreed that cherry made poor firewood because of its resinous character.

Other farmland trees are recorded in early surveys throughout the country, but usually in very small quantities. Holly was sporadically planted as a hedge tree in the north, although it was more usual as a feature of the wood-pastures there. It was rare in the south: on acid soils it occasionally accounted for as much as 3 per cent of the trees (as at London Colney in Hertfordshire: HALS D/EB 944 E1- 2 & P1), but it usually constituted considerably less than 1 per cent. There was, for example, only one holly among the 24,000 farmland trees recorded on the Earle estate in north-east Norfolk in 1722. Wild service was even rarer, recorded in tiny quantities at a handful of places, such as Navestock in Essex in 1722. Such trees had few uses and some of the examples recorded may have originated as thriving shoots in neglected hedges, left as pollards when most of their fellows were disciplined by renewed coppicing or plashing. Indeed, while we have emphasised throughout the importance of human agency in shaping tree populations, this involved the management of self-seeded trees as well as the active planting of new ones. In this context we should note how indigenous species such as maple, which were often present as shrubs in the kind of mixed hedges common in old-enclosed districts, could be found as pollards there, but only rarely in 'champion' areas. Walnut, in contrast, which *had* to be deliberately planted, was equally common in 'champion' and 'woodland' districts.

Trees of woodland and wood-pasture

The character of the trees growing in coppiced woods and wood-pastures was similarly shaped by the interaction of human choice and natural factors. The timber found within the former was, almost everywhere, overwhelmingly dominated by oak. Oaks thus accounted for all the 924 timber trees growing in Honeypot Wood, Wendling, Norfolk in 1777 (NRO HNR 135/3/18; NRO NAS1/1/14/ 7–9), all of those in a ten-acre wood in Finchingfield in Essex 1845 (ERO D/D Pg T8) and 95 per cent of those in woodland at Helmsley in Yorkshire in 1642 (Nrth ZEW IV 1/6). Again and again we find a contrast between the trees growing in woods and those on adjacent farmland, vividly illustrating the artificial character of both populations. Three examples from Essex make the point well. On the Topping Hall estate in Hatfield Peverel oak made up 48 per cent of the farmland trees in 1791, but accounted for all but one of the 2,000 trees growing in the Great Wood; at Finchingfield in 1773 oak constituted 57 per cent of the farmland timber but 100 per cent of the 968 trees in the four woods on the property; and at Little Baddow in 1777 it made up 65 per cent of the trees growing on the lands of Bicknaire Farm but 99.5 per cent of those in Bicknaire Wood (ERO D/DRa C4; ERO D/D Pg T8; ERO D/DRa C4). In Midland districts, where oak was often the second or third most important tree on farmland, the contrast was especially sharp. On the Montagu estates in Northamptonshire in 1749, for example, no less than 99.8 per cent of the 5,629 timber trees within Boughton Wood were oaks; at nearby Prist Coppice the figure was 99.9 per cent, as it was at Coppingford Wood (Boughton House archives). In part this importance reflects the primacy of oak as structural timber, already described. But the species' particular prominence in woods was also a consequence of the fact that it regenerates well in woodland conditions, although acorns were on occasions deliberately planted within existing woods (Wake and Webster 1971, 114).

More uncertainty surrounds the extent to which the coppiced understorey found in woods essentially represents the natural vegetation of areas in question, and to what extent, conversely, it was the consequence of accidental or deliberate management, or even of active planting. Ash, hazel and oak were the most important coppiced species in most areas, accompanied to varying extents by, in particular, elm, cherry, birch, beech, hornbeam, maple and (more rarely) lime, and with alder and willow on damp ground. There were clear, although under-researched, patterns of regional variation. On lime-rich boulder-clay soils, across East Anglia and much of the Midlands, woodland coppices were (and still are) usually dominated by ash and hazel, often accompanied by maple. On the clay-with-flints of the Chilterns, the North Downs and south-central England hazel tended to predominate over ash, and was sometimes accompanied by cherry. Oak tended to be

5.8 Coppiced stools of oak in Lea Wood, Matlock, Derbyshire. Oak coppice was more frequent in northern than in southern districts, probably for economic as much as for environmental reasons.

more important in the north, where it was accompanied by varying mixtures of hazel and ash, together with some rowan, alder, hawthorn, birch, crab, cherry and sycamore (Figure 5.8): Gledhill (1994, 245) has noted that oak, ash, elm and sycamore were all found in the woods at Downholme in the early nineteenth century, while 35 sycamore poles were recorded in a timber sale at Shipley in 1835 (Lds WYL639/339). Some types of coppice had more patchy, restricted distributions. Lime woods were and are concentrated on the area around the river Severn in Gloucestershire and Worcestershire, in Lincolnshire and on the Essex/Suffolk border. Woods in which hornbeam forms much or most of the understorey are found only in south Norfolk and north Suffolk, and in the area around London, the latter in particular mirroring its distribution in wood pastures. Sweet chestnut was largely, although not entirely, restricted to woods in Kent, Sussex and Surrey, and parts of the West Midlands. All these are very broad patterns, and there was and is much local variation. In Norfolk, for example, a handful of woods – such as Hockering Wood and Swanton Novers Great Wood – exist in which lime is a major component of the underwood, mainly in the centre and north of the county; oak coppice is found in a few places on acid, gravelly soils in the north of the county; Wayland Wood in Watton, and a few other woods in the central western part of the county, are characterised by coppice featuring large amounts of bird cherry (Barnes and Williamson 2015).

To a large extent the species present as underwood probably do represent the natural vegetation, modified by centuries of grazing before woods were enclosed and coppiced in the Middle Ages. The distributions of different coppice types tend to correlate, for the most part, with aspects of soil and climate. But there is no doubt that underwood was also weeded, managed, even deliberately replanted, and in this context it is important to emphasise that the three species most frequently coppiced in woods – ash, oak and hazel – produced material of particular practical and economic value. Hazel made excellent hurdles and provided hedging materials, wattles for the infill of timber-framed buildings, thatching broaches and firewood in the form of faggots. Ash poles were extremely useful, as we have noted, for fencing, hurdles, tool handles and minor building timber, as well as making particularly good firewood. Oak also made good firewood, as well as being the best fencing and building material, while the bark of the larger poles could be stripped and sold to tanners. Other important underwood species, of more local importance, are also distinguished by their utility. Hornbeam, for example, produces excellent charcoal and first-rate firewood. It would be surprising if foresters had not managed the understorey to encourage these particular species, perhaps removing less valuable shrubs that might compete with them, such as hawthorn or guelder rose. Contemporary comments often hint that natural developments towards the dominance of particular species were encouraged for profit: Pearce commented that in Berkshire woods the coppice was dominated by hazel, adding a little later that 'hazle-wood, in a county where great quantities of hurdles are wanting, is of course very profitable' (Pearce 1794, 54). But there is also some direct evidence for the deliberate modification of the underwood.

Many coppiced woods, as we have noted, were newly established or significantly extended in the course of the post-medieval period – for example, one on the Boughton estate in Northamptonshire was said in the 1760s to have been 'lately added to … and is not yet come into cutting' – and in such cases the coppice must have been deliberately planted with appropriate species (Boughton House archives). Where the archaeological evidence indicates that particular woods have expanded at the expense of farmland or common land the understorey of the extension is sometimes still significantly different to that in the original 'core'. The 'primary' portion of Wayland Wood in Norfolk, for example, has coppice characterised by a mixture of ash, hazel and bird cherry. That within the later extension to the east, probably added in the fifteenth century, mainly comprises, in contrast, hornbeam and ash (Barnes and Williamson 2015, 240–43) (Figure 5.9). But long-established coppices could also be augmented or replanted. A lease for South Haw wood in Wood Dalling in Norfolk, drawn up in 1612, bound the lessee to plant sallows in cleared spaces following felling (NRO BUL 2/3, 604X7); the tithe files of 1836 describe how there were 35 acres of coppice wood in Buckenham in the same county 'part of which has been newly planted with hazel' (IR 29/5816); while ash and sallow were planted in what appears to have been an existing wood near Boughton, in Northamptonshire, in 1752 (Boughton House archives). Lowe, writing about Nottinghamshire in 1794, noted the

5.9 Outgrown hornbeam coppice, with some ash, in the eastern section of Wayland Wood in Norfolk. Hornbeam dominates this section of the wood, which was probably added to the original 'primary' core – in which it is rare – in the fifteenth century.

importance of 'taking care of the underwood', described how 'vacancies are usually filled up with ash' and reported how, on one estate, the hazel and thorns were regularly stubbed up after the coppice was cut and 'and young ashes planted in their stead. By which mode … these woods have been very considerably improved' (Lowe 1794, 34, 114). Ash was the species perhaps most frequently propagated in woods. Rudge in 1807 describes how it was regularly replanted in the coppices in Gloucestershire and Vancouver in 1810 noted how, in Hampshire, some of the best ash shoots were retained when the coppice was felled and were layered 'in the vacant spaces' to form new plants (Vancouver 1810, 297). A similar practice is recorded in Surrey woods in 1809 (Stevenson 1809, 127).

Although modern writers have tended to emphasise the longevity of coppice stools (Rackham 1976, 14–15), eighteenth- and nineteenth-century writers and managers often noted their tendency to decline and decay, and the need for routine replacement. Boys suggested in 1805 that many of the coppices in Kent were regularly supplemented with new plants simply because 'wood, like everything else, decays and produces fewer poles every fall, unless they are replenished' (Boys 1805, 144). In 1815, at Ford Hill in Kent, one owner described how he was advised to fill up his existing coppices with 'chestnut and ash and willow plants to produce poles … and I found indeed it was highly needed as from the stubs dying from time to time which they are liable to do particularly willow stubs, large vacant spaces have got in many parts' (KHL U3750/21/1). One Herefordshire landowner described in 1852 how

> the wood after successive fallages deteriorate as numbers of the old stools die and unless there is a considerable amount laid out in filling up the vacant places with young wood, ditching, etc a quantity of useless stuff such as birch, orl [alder] and brambles grow up and consequently reduces the value of the wood. (HAR A63/111/56/12)

It is striking, in this context, that oak – common in coppices throughout the north and the west of the country – tended to dominate in the woods of mining areas, where it was, as we have seen, often managed on long rotations of 20 or more years. Indeed, such was the demand for pit props that oak was even sent from woods in Kent to the Newcastle collieries (Boys 1805, 141). But oak poles had other uses in an industrialising economy, providing good-quality charcoal and easily peeled bark for the tanning industry. It is hard to believe that oak was not deliberately encouraged as underwood in industrial areas. It is in this light, moreover, that we should perhaps view the local importance of other underwood species. Sweet chestnut was long established as a tree in England, although the attribution of its introduction to the Romans is, perhaps, less secure than often assumed. It was mainly a feature of woods in the south-east of England, where it was principally used for hop poles. These were required in vast quantities by the local hop fields – anything between 5,000 and 9,000 per hectare. Hops were first grown in England in the early sixteenth century, and poles of sweet chestnut, long and sturdy, were ideal for constructing the frames on which they grew. It is also interesting that the other main area of hop production, in the west Midlands, is the only other part of the country where the presence of sweet chestnut is well attested as coppice before the nineteenth century.

In most districts, wood-pastures were dominated by oak. This was particularly the case in the royal forests in the Midlands, where descriptions suggest that not only the timber but also a high proportion of the pollards were of this species. But in the south, and especially the south-east, beech was also a frequent tree both on commons and in deer parks. It was a particular feature of the Oxfordshire and Gloucestershire Cotswolds and also of the Forest of Dean, where there were said to be 'considerable quantities of fine large beech' in 1788 (Rudge 1807, 249). But the most extensive stands were in the Chiltern Hills of Oxfordshire, Buckinghamshire and Hertfordshire. William Ellis reported that beech was common along the whole length of the Chiltern Hills and their south-western continuation, from Dunstable as far as Wallingford (Ellis 1741, 25–6). But it was also abundant in wood-pastures on the long dipslope of the hills, especially in deer parks. In 1353, in order to obtain material for the pale surrounding the park attached to Berkhamsted Castle in Hertfordshire, the constable of the manor was obliged to sell beech trees to the value of £20, using the money to buy oak for the purpose – clearly suggesting that there were few examples of this species growing within the park (Whynbrow 1934). The nearby manor of Caddington contained a 'great beechwood of 300 acres', apparently a wood-pasture, in the early thirteenth century. There were 1,420 beech trees growing in the park at King's Langley in Hertfordshire in the mid-sixteenth century, far outnumbering the 300 ash and 180 oaks recorded (TNA E315/391), while large quantities were sold from the park at Knebworth in the early fifteenth century (HALS K117 and 119). By the eighteenth century, in both the Cotswolds and the Chilterns, enclosed woods of beech existed that were entirely composed of standards of different ages, without a coppiced understorey. Single trees or small groups were felled as required, the resultant spaces either filling naturally with self-sown specimens or being deliberately planted. Saplings were able to flourish because of the remarkable shade-tolerance of the species. These were not wood-pastures – livestock were carefully excluded – but many may have developed from them following the partial enclosure of areas of common land. It is noteworthy that Young emphasised the importance of making good fences between such woods and *adjacent* commons (Young 1809, 225). Others seem to have arisen as beech colonised local coppice woods, its shade gradually suppressing the oaks, or (more rarely) where beech was an element in the coppice. Ellis in 1741 described how it was often the case that, after the felling of the oak, 'a Wood of Beech has spontaneously succeeded; and when this has once got Dominion, it will be sure always to remain Master' (Ellis 1741, 26).

Beech was evidently well suited to the Chiltern and Cotswold soils, but Ellis believed that its importance in the former district was the consequence of human agency. Its suitability for the Chiltern soils had 'obliged our Fore-fathers, as well as those of the present Age, to set the Sides of their chalky Hills &c. with Beech-mast, where this tree will run up to a vast height with great Expedition' (Ellis 1741, 26). The huge numbers present in the park at King's Langley in the sixteenth century, at least, cannot have developed through the management of a natural, wild population. The park was not cut out of wooded 'wastes' but was created around 1290 at the expense of 'eight acres of meadow which used to be mowed before deer were placed therein … And one hundred and twenty acres of land which were arable', to which a further 160 acres 'of arable land' were added in 1397, and more in subsequent decades (Munby 1977, 152).

Beech was common as a wood-pasture tree in some other areas. It was a feature of Pamber Forest near Basingstoke, on similar soils to those found in the Chilterns; and, more importantly, it was frequent on the poor soils, formed in London clays and pebble gravels, lying to the north of London and, to a lesser extent, to the south. Beech worth 68 shillings was thus sold from the Great and Little Parks at Hatfield in south Hertfordshire in 1428 and a combined total of 12,000 oak and beech trees was recorded there in a survey of 1538 (HHA Manor Papers: Summaries I, 339; Page 1902, 99). A further survey, in 1626, recorded twice as many beech (520 trees) as oak (227) in Hatfield Great Wood, the former Great Park, while large amounts of beech trees were sold from Hatfield New Park, including 400 in 1629 alone (Austin 1995, 6–7; HHA Sals. General 44/1 and Sals. Legal 66/7). To the south of Hatfield beech comprised about 20 per cent of the trees growing in the woods – probably wood-pastures – on the manor of North Mymms in the early fifteenth century and it was a major component (together with oak and hornbeam) of the woods in Broxbourne and Hoddesdon (HALS 80226; Page 1912, 430). Ancient beech pollards still survive in large numbers in Epping Forest in south-west Essex, accompanied by hornbeam (above, Figure 4.6), and at Burnham Beeches, north of Slough, where they are accompanied by oak.

Beech was generally accounted a useful timber by seventeenth- and eighteenth-century writers. It was used to make floor boards, external weather-boards, furniture and threshing floors, while its resistance to decay in waterlogged conditions meant it was much sought after for the construction of mills and ships. Mortimer thought that it was good for making trenchers, shovels and spades, and noted its excellence as firewood (Mortimer 1707, 338–9). None of this explains its particular association with wood-pastures, whether common woods or private deer parks. Ellis, however, noted that 'The Beech excels all other Trees in Parks &c. for the Returns it makes of prodigious Quantities of sweet, healthful Mast, which greatly helps to subsist the red and fallow Deer' (Ellis 1741, 41). Moreover, beech is equal only to hornbeam and oak in its ability to tolerate intensive browsing (Rackham 2006). Whether it was planted in wood-pastures, or was simply able to colonise and survive there better than other species, remains a moot point.

The other great wood-pasture tree of the south-east was hornbeam. This was usually the dominant tree on the extensive commons found on the poor soils around London, where it stood in marked contrast to the oak, elm and ash that dominated the surrounding hedgerows. At Drakeshill Farm, in Navestock, Essex, in 1722 hornbeam made up 15 per cent of the 442 mature trees growing in the fields and hedges, but 85 per cent of the 959 growing on the adjacent area of common land, 'Slade Common Wood' (ERO D/DU 583/2). It could also be found, often in the company of oak and beech, in small wood-pastures on private land in this same district. A note on a map of a property in Woodford in Essex, surveyed in 1772, described how 'The ground … is almost covered with beech hornbeam and oak pollards which are about a foot diameter 3 feet from the ground' (ERO T/M 149/1). Hornbeam pollards existed in large numbers on the commons to the south to the Thames, especially on Ashtead Common and Wimbledon and Putney Commons, well into the nineteenth century (Howkins and Sampson 2000). They were also a characteristic tree of forests and deer parks, not only in the area around London (as far north as Hatfield Forest, close to Stansted airport, where fine examples remain) but also in north Suffolk and south Norfolk (Figure 4.5). A valuation of the Suffolk estates of Bury St Edmunds Abbey, made at the Dissolution in 1539, described how the park at Redgrave contained '30 acres thyn sett with pollyng oakes and hardebeme growing by parcells' (Bury St Edmunds RO Accn. 1066). As already noted, in both these districts hornbeam was also the dominant species in the understorey of many coppiced woods.

Hornbeam was poorly represented in the prehistoric landscape of these same areas, to judge from pollen evidence. Pollen cores taken from waterlogged deposits within Epping

Forest show that it first appeared in the area during the Saxon period, together with beech; but both species became prominent only some time later, after the decline of small-leafed lime, a development that was itself initially followed by an increase in birch and hazel (Baker *et al.* 1978). The later phases of this pollen sequence were not radiometrically dated, but could suggest that the classic beech/hornbeam vegetation of the Forest developed as late as the twelfth or even the thirteenth century. The pollen sequence from Diss Mere indicates that hornbeam was present in the south Norfolk/north Suffolk area from the late Iron Age or Roman periods, but only at very low frequencies. It became a significant component of the local vegetation some time in the medieval period, the precise date being uncertain as, once again, the pollen sequence is itself not radiometrically dated (Peglar *et al.* 1989). It is noteworthy that while oak, ash, elm, hazel, beech and maple all occur as elements in pre-Conquest placenames, hornbeam does not: probably the earliest example is Hornbeamgate, a hamlet in Hatfield in Hertfordshire, first recorded as late as 1366 (Mawer and Stenton 1938, 224). Hornbeam does not seem to feature much in documents before the second half of the fourteenth century, moreover, when it is referred to in the accounts for a number of deer parks (Rowe and Williamson 2013b, 156–9). In the fifteenth century the bailiffs of parks at Great Munden and Knebworth recorded receipts for 'trees called hornbeam', as though the name, and the tree itself, were unfamiliar to them (Rowe 2009, 34).

Hornbeam thus appears to have become widely established in wood-pastures only during the Middle Ages, and there are indications that its importance increased still further thereafter. In 1538 Hatfield Great Park in Hertfordshire was said to contain 2,000 oak and beech, but by 1626 it contained 5,227 hornbeams 'of all sorts', ten times the number of beech, the next most numerous species (Austin 2013, 143). By this time, huge numbers were present on the local commons, such as the 24,000 recorded on Cheshunt Common in 1695 (Rowe and Williamson 2013, 156–7). It is hard to account for hornbeam's rather late rise to prominence unless it was actively planted or otherwise encouraged, in both woods and wood-pastures, and Anne Rowe has recently argued that it was being deliberately established on common land as late as the seventeenth century by manorial lords (Rowe 2015, 310–14). It is certainly clear that some private wood-pastures of hornbeam were created at a remarkably late date. The old hornbeam pollards that are such a feature of Hatfield Forest in Essex (Figure 4.5, above) were established, at the expense of enclosed coppices, in the years around 1700 (Rackham 1989, 6). The importance of hornbeam as a coppice in woodland in both the London area and East Anglia may also have been – in part at least – a consequence of deliberate planting. Indeed, in 1742 William Ellis commented on how, in Hertfordshire, '[t]his [the hornbeam] is in great reputation for both copsehedge and wood and is planted in many parts, but more abundantly about Whethamsted [Wheathampstead] in this County' (Ellis 1741, 71). As well as having good resistance to grazing, the wood of hornbeam had a range of specialised uses that reflect its name – the 'hard beam'. Langley thought it of 'great Use to the Mill-wright, for Coggs to his Wheels, as well as to the Turner, Carpenter and Joiner'. But he added that 'for Fire-wood there's none better' (Langley 1728, 161–2). Ellis similarly considered that it 'so far excels most other Fire-woods, that when it is burnt enough, the Coals will hold a bright Fire like Charcoal for a long time' (Ellis 1741, 72). It also, and perhaps more importantly, made excellent charcoal. It may be no coincidence that the two main concentrations of hornbeam in England – in south Norfolk/north Suffolk, and around London – lie close to the country's two largest medieval cities, Norwich and London, both hungry for fuel (Barnes and Williamson 2015, 80–85).

Lastly, we should note the existence of specialised woodland and wood-pasture trees in the north of England, where the practice of feeding livestock on leaf fodder continued in some districts into the nineteenth century. We saw earlier how ash was widely planted, in part for winter feed, both on enclosed land and, apparently, on commons. In many northern areas holly was also used in this way (Spray 1981). Abraham de la Pryme described in 1696 how 'in south-west Yorkshire at and about Bradfield and in Derbyshire they feed all their sheep in winter with holly leaves and bark ... To every farm there is so many holly trees ... care is taken to plant great numbers of them in farms hereabouts' (Jackson 1870). At Newburgh on the Howardian Hills in Yorkshire a survey of 1605 shows a high concentration of trees within a knot of small closes just to the north of the deer park with names containing references to 'haggs' of 'hollin' or holly (Nrth ZDV); 'haggs' were frequently found within deer parks, where the holly was cut for deer feed (Jones 2012, 37). A survey of the Manor of Sheffield made in 1637 refers to 27 'Hollin haggs'. But holly was also planted, or encouraged, on common land. In 1725 a party travelling across Birley Moor and Hollinsend, near Sheffield, rode through

> the greatest number of wild stunted holly trees that I ever saw together. They extend themselves on the common for a considerable way. The tract of ground they grow upon is called the Burley Hollins ... . [They have] their branches lopped off every winter for the support of the sheep which browse upon them, and at the same time are sheltered by the stunted part that is left standing. (Jones 2012, 135)

Similar holly-dominated wood-pastures could be found in the Lake District. Pennant, travelling near Hawkshead in Cumbria in 1772, 'in one place observed a Hollypark, a tract preserved entirely for sheep' (Pennant 1776, 36), while Thomas West described how in Furness holly trees were 'carefully preserved' for fodder 'where all other wood is cleared off, and large tracts of common pasture are so covered with these trees, as to have the appearance of a forest of hollies' (West 1774, xlv). As late as 1794 the agent of the Bolton estate said that 'a very great havock was made last winter amongst the Hollies, which ... were cutt down for the purpose of fodder for Cattle', suggesting that in some places holly continued to be employed as an emergency fodder crop throughout the eighteenth century (Dormor 2002, 197).

Conclusion

There seems little doubt that, while in part the consequence of variations in soil and climate, the particular balance of tree species found on farmland and in woods and wood-pastures was critically shaped by human agency: either directly, through choices influenced by sound economic and practical considerations, or indirectly, as a side-effect of other aspects of land management. Although, as we have emphasised, the available sources need to be interpreted with caution, there is no real doubt that oak, ash and elm were generally the dominant farmland trees, albeit with a relative importance that varied from region to region. They were undemanding in their requirements and useful in a wide variety of ways. Other trees were less useful, more demanding, or were better grown as coppice. They accordingly occur – at least as *trees* – at low levels in our records, likewise with regional preferences. The artificial character of rural tree populations is perhaps particularly clear in the case of maple, which seeds freely, is widespread in hedges and in the understorey of woods, but which was seldom allowed to grow as a standard or pollard, although it makes a fine tree. Only in particular circumstances did trees other than oak, ash and elm dominate the landscape, such as the willows in alluvial wetlands and meadows, especially in the Midlands, or the hedgerow fruit trees found in various districts.

The situation with woods and wood-pastures was more complex. Beech and hornbeam were important in many southern wood-pastures because they had good resistance to grazing and provided browse for stock, as well as other economic benefits, but how far their dominance in particular places was a long-term side-effect of their use for grazing or how far they were deliberately planted in such contexts because of these characteristics remains unclear. In woodland, oak was the main timber tree as a consequence of economic choices, coupled with its tolerance of shade. It is usually assumed, in contrast, that the coppiced underwood was largely 'natural' in character, its composition inherited from the wild vegetation or at least a consequence of soils and drainage, but as we have seen this, too, was heavily modified by human intervention and the coppiced understorey was sometimes deliberately replanted.

In short, while human choices about which species to plant, preserve or encourage in particular contexts were contingent on environmental factors – on soils, climate and drainage – the composition of tree populations in the post-medieval countryside was nevertheless highly unnatural in character, if by 'natural' we mean largely uninfluenced by human actions. The almost complete absence of small-leafed lime from farmed landscapes, woods and wood-pastures by the start of the period studied here is a striking testimony to the true character of rural tree populations.

 CHAPTER SIX

Trees in the modern world, c.1780–1880

Contexts: the revolutions in agriculture and industry

As we have emphasised on a number of occasions, the character of woods and farmland trees recorded in the seventeenth and eighteenth centuries, when documentary evidence becomes particularly abundant, does not necessarily provide a good guide to the situation in 1500, still less in 1300. Trees and woodlands were embedded in wider patterns and systems – economic, technological and social – that were themselves constantly changing. This said, it seems likely that the period from the later eighteenth century to the end of the nineteenth saw more radical changes in their character than anything that had occurred for many centuries, with the development of novel forms of management, the complete demise of many long-established practices and environments and the widespread planting of new species introduced from abroad. It is also true, as we shall see, that in some ways there was more continuity in management systems – or, at least, a slower rate of change – than is often suggested. But it is the sheer scale of transformation that is the most striking feature of the period.

The ultimate causes of these profound changes are complex and cannot be dealt with in any detail here: in the last analysis, they were intimately associated with the birth of the modern world. The early eighteenth century had witnessed the gradual emergence of a truly global system of trade, a development in which Britain, with its new colonies in the Americas, was centrally involved (Colley 2009, 67–71). But from around 1760, and especially from the 1780s, economic growth stepped up a gear. The population of England began to increase rapidly, rising with unprecedented speed from around five and half million in 1750 to some nine million by 1800, and to nearly 19 million by 1861 (Figure 6.1) (Mitchell and Deane 1962, 5–6). This engendered a series of profound changes in farming: the classic 'agricultural revolution', in which the widespread adoption of a range of new crops, rotations and techniques boosted farm production. Even in 1851 imports accounted for only around 16 per cent of food consumed in England and Wales. To feed an expanding population the volume of wheat produced in the country more than doubled, while that of barley may have increased by over two-thirds, in the course of the later eighteenth and early nineteenth centuries, and the growth in the production of other foodstuffs was on a similar scale (Holderness 1989; Beckett 1990, 9).

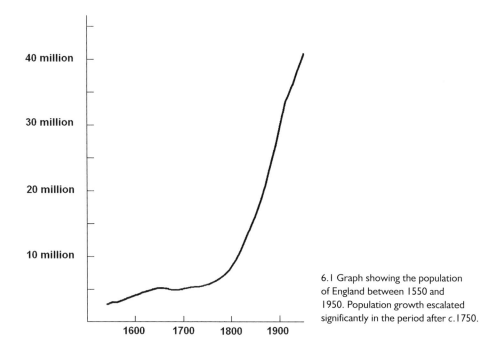

6.1 Graph showing the population of England between 1550 and 1950. Population growth escalated significantly in the period after c.1750.

The rising demand for food, coupled with a fashionable concern for agricultural 'improvement', had profound effects on the way people regarded a farmed landscape that was, in many districts, liberally cluttered with trees. But agricultural modernisation impacted in other key ways on trees and woodlands. Firstly, it intensified existing and long-established trends towards the concentration of landed property in the hands of large estates, especially on the more marginal soils, where smaller proprietors often lacked the capital now required for successful farming. Many factors contributed, and had long been contributing, to the consolidation of properties in ever larger units, not least the general expansion of world trade in the eighteenth century, the benefits of which passed disproportionately to major landowners. Such matters cannot be discussed in detail here (Clemenson 1982; Williamson 2007; Beckett 1984; 1986), but what is clear is that the consolidation of land ownership encouraged a wave of replanting, for large proprietors could afford the luxury of investing in forestry – an enterprise of uncertain and long-term benefits – in a way that smaller ones could not. Many of the newly planted areas were former common land, moreover, for a key element in the agricultural revolution was the enclosure, usually by parliamentary act, both of the remaining open fields and of surviving tracts of common grazing, including wood-pastures and forests. Where not converted to farmland, poor ground, especially in the form of heaths and moors, was often seen as a good place to plant trees.

The 'agricultural revolution' not only served to feed a rapidly growing population but also brought about a dramatic improvement in what economic historians call 'labour

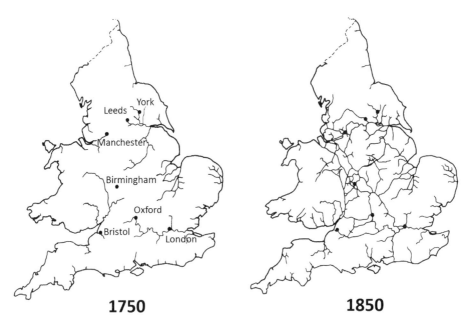

6.2 The development of navigable waterways (above) and railways (below) in England during the period of the industrial revolution.

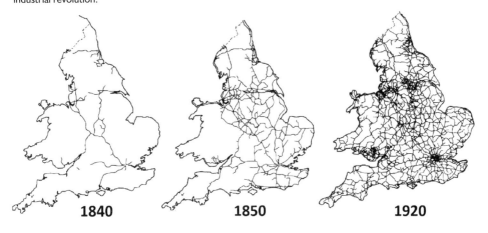

productivity': that is, the number of individuals required to produce a given amount of food (Overton 1996, 121–8). In 1760 the output of each agricultural worker could feed around one other person, but by 1841 it could feed another 2.7. This was vital because it was in this period that England also, and perhaps more importantly, became a truly industrial economy. As we have seen, already – by the start of the seventeenth century – there were many thriving industries, some of which had a profound impact on the management of trees and woods in their immediate locality. But the scale of industry now increased almost exponentially. The output of British coalmines rose steadily from 2.5 million tons in 1700 to around 8 million by 1780, but then increased more rapidly, reaching 11 million tons by 1800 and 22 million by 1830 (Flinn 1984, 27). This exploitation of the country's 'underground forests' fuelled other industries. There were fewer than 20 blast furnaces in England in 1700; by 1805 there were 177 and by 1852 at least 665, all now fired with coke rather than charcoal. In 1700 around 30,000 tons of pig iron were produced in England each year. By 1850 the figure was two million (Cossons 1987, 34). The domestic consumption of coal also increased rapidly from the late eighteenth century. In 1790 only half a dozen counties, mainly located on coal fields, had been largely or entirely coal-burning; by the 1850s and 1860s it is likely that all were.

Central to industrial expansion, and especially to the spread of coal as the principal domestic and industrial fuel, was a steady and profound improvement in transport infrastructure. Of particular importance was the development of a national canal network, something that began in earnest with the completion of the Bridgewater canal (designed to serve the coal mines at Worsley) in 1761 and reached 'a crescendo in the "mania" of 1789–93' (Cossons 1987, 256) (Figure 6.2). This was directly paralleled by the expansion in the numbers of turnpike trusts, which was, likewise, most rapid in the period between 1750 and 1850 (Bogart 2005; Albert 1972). All this allowed coal to be transported much more easily and cheaply, and its use now spread far beyond coalfields, coasts and major rivers. This said, in many areas it was only the arrival of railways in the first half of the nineteenth century that sounded the final death knell for traditional forms of firing: by 1850 nearly 7,000 miles of rail line had been built, and in all parts of England, except the west, few places were more than 15 miles from a station (Figure 6.2). On the Bedfordshire brickfields, for example, coal became the main form of fuel only following the arrival of the local rail lines in the 1840s, the Duke of Bedford's steward, Thomas Bennett, recalling in 1869 how 'Furze used to be grown for a demand for brickmaking, but this fell off some years ago' (Cox 1979, 44).

Improved systems of transport, economic expansion and an increasingly interconnected world had other effects. Timber had long been imported into England, especially from the Baltic. As early as 1273 an entry in the Account Rolls of Norwich Cathedral Chapter recorded the journey of 'John the carpenter' to Hamburg to buy timber and the carriage of boards from Hamburg to Yarmouth (Latham 1957, 28). By the post-medieval period such references are common. Baltic oak was brought to Collyweston in Northamptonshire via Boston and Peterborough in 1500, for example, while in 1550 John Johnson was importing oak from Antwerp to King's Lynn, and then transporting it by water to Yaxley and on by land to Glapthorn in Northamptonshire (St John's College Cambridge, MS 102.9). The eighteenth century saw a rapid increase in the scale of imports, however, especially into east coast ports. Norway was a particularly big exporter because its saw mills, equipped with fine blades, were thought to produce the best quality deals. Large quantities came into northern England via the port of Hull (Davies 1964) and during the eighteenth century the development of sawmills both in and around that town meant that there was an increase in the volume of imports of unsawn rough wood 'from nothing at the beginning of the century to 1,135 loads in 1758' (Jackson 1975, 16). The quantity of timber passing through the port increased to 8,260 loads in 1768 and to nearly 21,000 loads by 1790. By the end of the eighteenth century 'the new mills, foundries, potteries and workshops east of the Pennines were constructed from wood and iron imported through Hull' (Jackson 1975, 21–2).

The same period also saw an increased dependence on the New World as a timber source – initially for specialised woods such as mahogany, but by the nineteenth century for a wide range of general hardwoods and conifers. All this meant that England no longer needed to be self-sufficient in timber and many agricultural improvers, looking at Britain's expanding industries, teeming population and urgent need for food, believed that it would be better to use the space taken up by woods and trees for growing crops or grazing livestock. The enquiry held by the House of Commons into the state of timber reserves in 1790, motivated by perennial concerns about the supply of oak timber to the Royal Navy, heard a number of comments on these lines:

> It is the Improvement of the Kingdom, a Thousand Times more valuable than any Timber can be, that has wrought the very good and proper Diminution of Oak and it is to be hoped the Diminution will continue … . While we are forced to feed our People with Foreign Wheat, and our Horses with foreign Oats, can raising Oak be an Object? The scarcity of Timber ought never to be regretted, for it is a certain Proof of National Improvement: and for Royal Navies, Countries yet barbarous are the right and proper Nurseries. (*Journal of the House of Commons* 1792, 343)

Another respondent believed, more specifically, that 'There can be no doubt that *Russia, Sweden, Norway*, and *America*, which abound with immense Forests, will at all times supply us with Timber, and take our Manufactures in Barter' (*Journal of the House of Commons* 1792, 342). In short, industrialisation, globalisation, agricultural modernisation and changes in land ownership – all interconnected in many and complex ways – together served to revolutionise the character of England's trees and woodlands in the century after *c.*1780.

Farmland and the fuel revolution

We have already noted how, in coal-mining districts and other areas where abundant supplies of alternative fuels, such as furze or peat, were available, there were already relatively few hedgerow trees by the eighteenth century. As canals proliferated and roads improved, coal could be transported economically into more and more distant areas, and at a time when farmers and landowners were keen to increase the scale of agricultural production. Not only did high densities of farmland trees now appear an unnecessary impediment to farming but so too did the dense mesh of hedges that existed in many districts, and which had in part existed in order to provide fuel. Small and irregularly shaped fields were difficult to plough, tall spreading hedges and an abundance of hedgerow trees shaded out the crop and robbed the soil of nutrients, and wide 'hedge rows' took up substantial areas of potentially productive land. Already, in 1790, respondents to the government enquiry into timber supplies were able to describe how in almost all districts the increasing use of coal was having a significant impact on farmland trees. The removal of timber and hedges was 'frequent', 'becoming common' or the 'general practice' in many counties. This was especially the case in parts of East Anglia and the south-east, where the area under arable cultivation was expanding fast at the expense of permanent pasture. One contributor typically noted that 'The county of Hertfordshire consists chiefly of Land in Tillage, and by … clearing the Hedges of all kinds of Trees they admit of plowing to the utmost Bounds of their Land' (*Journal of the House of Commons* 1792, 318).

Local records present a similar picture. The rector of Rayne in Essex observed in his tithe accounts of the 1790s how on one farm in his parish 'the fields were over-run with wood', but 'since Mr Rolfe has purchased them, he has improved them by grubbing up the hedgerows and laying the fields together' (Williamson 2002, 95). In 1801 one observer of the Essex countryside was able to declare: 'what immense quantities of timber have fallen before the axe and mattock to make way for corn' (Brown 1996, 34). Randall Burroughes, a gentleman farmer in Wymondham, on the Norfolk boulder clays, was at the forefront of this war on the 'traditional' landscape, the details of which are carefully recorded in his farming journal from the 1790s. Every winter his men were busy stubbing out hedges and taking down trees. In the last two weeks of 1794, for example, Burroughes described how 'Elmer & Meadows began to through down & level an old bank in part of the pasture between little Bones & Maids Yards'; reported that 'the men were employ'd in stubbing a tree or two for firing & other odd jobs'; described how 'some ash timber' was cut down; and noted how 'the frost continued very severe so much so that … the men employed in throwing down old hedgerows found the greatest difficulty in penetrating the ground with pick axes' (Wade Martins and Williamson 1995, 48–9). Throughout southern and eastern England wide, spreading hedges were hacked out and replaced with others, straighter and neater in form, or were removed entirely to create larger enclosures.

Pollards were a particular target for the improvers. One respondent to the 1790 enquiry reported succinctly of Suffolk: 'Old Pollards, which furnished Fuel, being taken down'. Anthony Collet described in the 1790s how Suffolk landlords were giving their clayland tenants leave to 'take down every pollard tree that stands in the way of the plough' (Young 1797, 57) The landed elite had always been concerned about the dominance of pollards over timber in the hedgerows of their estates and about the proclivity of tenants to convert the latter into the former. Such antipathy now reached obsessive proportions (Petit and Watkins 2003). In 1796 William Marshall asserted that:

> We declare ourselves enemies to Pollards; they are unsightly; they encumber and destroy the Hedge they stand in (especially those whose stems are short), and occupy spaces which might, in general, be better filled by timber trees; and, at present, it seems to be the prevailing fashion to clear them away. (Marshall 1796, 100–101)

Apart from any rational economic objections, pollards were looked on by improvers as relics of backward peasant agriculture, and they may, additionally, have offended 'polite' taste because they were as 'unnatural' as the topiaried trees in formal gardens, which were now out of fashion among the rich and educated (Thomas 1983, 220–21). 'Let the axe fall with undistinguished severity on all these mutilated heads', urged Thomas Ruggles in 1796 in the *Annals of Agriculture* (Ruggles 1786, 180). But it is the change in local fuel economies that mainly explains the rapid reduction in their numbers.

Estate accounts often record the sale of many hundreds of pollards, ruthlessly culled from hedges, in the early decades of the nineteenth century, as, for example, in Earsham or Bylaugh in Norfolk (NRO HNR 149/2; MEA 3/549, 658). In 1822 the steward of the Broome estate in north Suffolk suggested to the owner, Sir William Fowle Midleton, that:

> A great many old pollards may be taken down … these of course will give little or no bark and are only fit for charcoal burners and for firewood in general, but if you wish to have any pollards felled to clear some of the lands and pastures, which would indeed be highly necessary and of great benefit to the tenants, it is quite another thing. (Ipswich RO HA 93/3/720)

We have seen how eighteenth- and early nineteenth-century surveys from old-enclosed areas commonly show more than two-thirds of the trees as pollards. By the middle of the nineteenth century figures were significantly lower. On an unnamed estate in north Norfolk in 1835 only 38 per cent of the trees were pollards; of the 1,041 trees described in a survey of the Evans-Lombe estate in the centre of that county in 1850 only 20 per cent were so described (NRO HNR 112/19). This said, the speed with which old-enclosed landscapes were tidied up and modernised was uneven. Small proprietors might still establish new pollards in their hedges and, in many old-enclosed counties, especially those lying at a distance from navigable waterways, agricultural improvers continued to rail against large hedges, small fields and an abundance of pollards well into the nineteenth century. In 1849 William and Hugh Raynbird could still describe how,

> Like other early enclosed counties, a great part of Suffolk is disfigured in all directions with hedges and ditches; many of these might be removed without injury to the drainage. The removal of the hedges that are not required, and the pollard-trees with which so many of them are so thickly studded, would reclaim more waste than the bringing the tracts of heath into cultivation; for these are worse than waste; they require an annual expenditure to keep them in repair; their roots … injure the crops to a great distance into the field; their shade delays the ripening and the harvesting of the crops, and harbours an infinite variety of vermin in the shape of birds and insects. (Raynbird and Raynbird 1849, 62–3)

The various contributors to the Raynbirds' volume on Suffolk farming estimated that between 6 and 10 per cent of the land area in Suffolk was lost to hedges and lines of pollards

(Raynbird and Raynbird 1849, 121–2, 145). Many insisted on the necessity of reducing the size of hedges and 'cutting down the still remaining pollard trees and timber'. Auctions of redundant pollards continued in many districts through the middle decades of the century; for example, pollards made up 96 per cent of the 389 trees sold at Tingate Farm, Broxted in 1852, most of the others being 'spars' presumably thinned from hedges (ERO D/F 35/3/135). By the end of the century pollards are seldom mentioned in documents relating to farmland trees. The pollarding of hedgerow trees continued well into the twentieth century in many rural districts, but it was now generally on a sporadic, casual basis.

Measuring tree loss

Given that in most districts the majority of hedgerow trees had been pollards, it is not surprising that the late eighteenth and nineteenth centuries saw a marked reduction in the density of farmland trees. On the East Anglian boulder clays in particular tree numbers fell by around 80 per cent in the course of the later eighteenth and the nineteenth centuries, from over 25 per hectare in the 1770s to fewer than 5 by the 1880s. On the loams of north-east Norfolk the scale of the decline was similar, from an average of 20 per hectare to

6.3 The area around Beeston-next-Mileham, Norfolk, as depicted on the first edition Ordnance Survey 6-inch map of the 1880s. The phenomenal density of trees mapped by Henry Keymer in the previous century has been drastically thinned (see Figures 2.1 and 3.1)

around 3. These late nineteenth-century figures are derived from the first edition Ordnance Survey 25-inch maps, surveyed in the 1880s, which may underestimate tree numbers to some extent (omitting examples where their depiction interfered with other detail, for example). But the broad scale of the change suggested is probably reliable, and at local level the reductions might be even greater. The 531 trees mapped by Henry Keymer on a property at Beeston-next-Mileham in Norfolk in 1764 (NRO WIS 138) had been reduced, by 1884, to around 80 (Figure 6.3); the 1,096 trees recorded on a farm in Scarning by the same surveyor in 1761 (NRO BCH 20) had by 1884 fallen to a mere 63. To some extent these reductions were associated with an increase in the size of fields and a concomitant reduction in the length of boundaries. On the Beeston property mapped by Keymer, for example (NRO WIS 138), average field size rose from c.1.4 to 3.6 hectares: indeed, the field pattern here was substantially redrawn between the two dates, the irregular pattern of enclosures partly replaced by a new, rectilinear arrangement reminiscent of the landscapes created by parliamentary enclosure. Indeed, the removal of hedgerow trees was often accompanied, in old-enclosed districts, by the rationalisation of field patterns, field amalgamation and the replanting of hedges. But even where there was little or no change in field size, reductions could be massive. At Scarning, for example, there was only a modest increase in average enclosure size, from 1.4 to 2 hectares, yet tree densities fell by nearly 85 per cent. Changes in other old-enclosed districts, especially in the east of England, were on a similar scale. In Hertfordshire, average densities on the boulder clays in the north-east of the county appear to have fallen from around 25 per hectare in the later eighteenth and early nineteenth centuries to around 3.5 per hectare by c.1880, and on the London clays in the south from 25 to around 4.5. In the neighbouring counties of Essex and Suffolk reductions were on a comparable scale.

Where Midland 'champion' parishes had been enclosed in the seventeenth or early eighteenth century, before the advent of parliamentary enclosure acts, there was also usually a significant loss of trees through the later eighteenth and nineteenth centuries. The Northamptonshire parish of Armston, for example, was already fully enclosed in 1749, when a survey recorded 1,721 trees there (Boughton House archives). By the 1880s there were 738 – a decline of 57 per cent. The density of 10.6 trees per hectare on John Darker's scattered estates in Northamptonshire in 1791 had fallen to around 2.5 by the 1880s, while the 9.6 per hectare on the Duke of Powis' properties in Upper and Nether Heyford, Glassthorpe and Newbold in 1758 declined to around 2.6 by the same date (NHRO ZB 1837; NHRO YZ 2183) – a return, in effect, to pre-enclosure levels. Sometimes such declines were a consequence of the removal of free-standing timber from pasture closes; sometimes they were related to the reduction in the length and numbers of field boundaries; but often, as in the Northamptonshire parishes of Elkington and Nobottle, field patterns remained largely unchanged, and the loss was simply a consequence of a reduction in the numbers of trees present in hedges. Such declines in tree numbers were not universal. In a few places, mainly in the vicinity of large country houses, numbers held up well, or even increased – a phenomenon that was, indeed, repeated to an extent in most parts of the country. The 1749 timber survey of Warkton, near the heart of the Boughton estate and next to Boughton Park itself, lists 1,350 trees, whereas the Ordnance Survey shows nearly 2,000, although a proportion of these stood in avenues. But, on the whole, in Midland parishes enclosed before the late eighteenth century tree numbers fell significantly thereafter.

To some extent the scale of this decline is surprising given that already, by c.1780, access to coal from the mines of Warwickshire, Staffordshire and Nottinghamshire had ensured that the numbers of pollards were much lower than in old-enclosed, woodland areas. The explanation is, in part, that the period also saw a growing antipathy on the part of many farmers and landowners to having *any* trees in hedges, timber or pollard. They were all 'incumbrances' that needed to be removed: 'the farmers say "knock them up, no corn will grow near them"' (Tuke 1800, 196). Timber should be grown in woods or plantations, or imported from the vast forests of the Baltic: farmland was for growing food. This explains why, in some but by no means all districts, the new hedges created by parliamentary enclosures, and especially those planted after c.1800, were singularly lacking in trees. In Rutland, for example, it was noted in 1808 that 'there is in general a great neglect in the late enclosures of planting timber in the hedgerows'; while in Nottinghamshire in 1794 'It is to be lamented, that in the new inclosures very little attention should have been paid to raising hedge-row timber' (Lowe 1794, 57). The Northamptonshire parish of Luddington lay almost entirely open until its enclosure by parliamentary act in 1807, and in 1749 contained around 517 trees (Boughton House archives); the first edition Ordnance Survey map from the 1880s depicts only 634, even though the parish was now filled with hedges.

An increasing hostility towards farmland trees of any kind is also apparent in most northern counties. Even where coal had long been the major fuel and pollards few, trees likewise declined significantly, although here perhaps at a slightly later date. The first Edition Ordnance Survey 6-inch maps for the north of England were surveyed earlier than in the Midlands and south, in the 1850s, and this often allows us to chart the chronology of decline with more precision. At Sinderby in the Vale of Mowbray in Yorkshire a map of 1778 suggests a density of around 0.6 trees per hectare, which had increased significantly, to 1.7 per hectare, by the 1850s; but by the end of the century it had apparently fallen back to a mere 1 per hectare (Nrth ZIQ). At Skeeby, also in the Vale of Mowbray, the 0.5 trees per hectare apparently recorded on a map of 1779 had risen to around 2.9 by the 1850s, but had fallen back to 1.2 per hectare by the 1890s (Nrth ZMI). In some townships losses in the late nineteenth century were on a huge scale: at Birdforth on the edge of the Hambleton Hills 75 per cent of the trees shown on the 6-inch map of the 1850s had disappeared by the end of the century. Some upland parishes, especially in Yorkshire, show a similar pattern: for example, at Brandsby in the Howardian Hills the density of 0.8 per hectare indicated by a map of 1746 (Nrth ZQG IV 1/1) had risen to over 2 by the mid-nineteenth century before declining to c.1.5; and at Beadlam, on the margins of the North York Moors, a map of 1785 (Nrth ZEW M 13) suggests a density of just under 0.9 trees per hectare, which rose

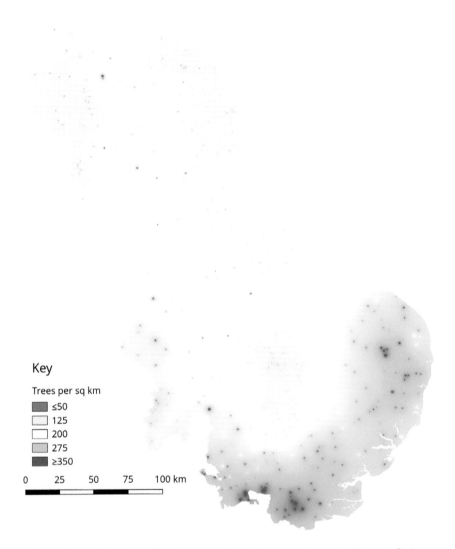

Key

Trees per sq km

- ≤50
- 125
- 200
- 275
- ≥350

0   25   50   75   100 km

6.4 'Heat' map, showing the density of hedgerow trees depicted on the late nineteenth-century Ordnance Survey maps in eastern England. The distinction between old-enclosed 'woodland' districts and late-enclosed former champion ones is clear (compare with Figure 1.7).

to around 2.5 per hectare in the 1850s before declining to around 1.6 by the end of the century. At nearby Pockley, a seventeenth-century survey (Nrth ZEW IV 6/4/6) suggests a density of 0.6 trees per hectare, while a map of the same township made in 1785 (Nrth ZEW M 14a) suggests 1.1 per hectare: this had risen to 3 per hectare by the 1850s before falling back, in the manner by now familiar, to 2.2 by the 1890s. It seems likely that the initial modest expansion in tree numbers in these places was the consequence of a rising demand for timber, fuelled by industrialisation and urbanisation, but this was steadily undermined by the rising tide of imports and prevailing fashions among landowners and farmers, so that, as trees were felled, there was a growing reluctance to replace them.

By the last decades of the nineteenth century significant regional variations in the numbers of farmland trees remained. This was in part because their density had been so great in many old-enclosed, 'woodland' districts lying at a distance from the main coalfields that, even after nearly a century of attrition, they remained numerous. But in part it was because late enclosures, for the reasons we have explained, often created sparsely timbered hedgerows. As a consequence, the old division between 'woodland' and 'champion' was ghosted, as it were, in the densities of farmland trees shown on the Ordnance Survey 6-inch maps from the 1880s and 1890s (Figure 6.4).

While a number of influences shaped the density of farmland trees, changes in fuel consumption and the rising scale of timber imports were probably the most important. The former also had a determining influence on the character of hedges planted in the period after c.1780, whether as a consequence of parliamentary enclosures or as field patterns were rationalised in old-enclosed districts. Although some small proprietors continued to plant mixed-species hedges, the general practice was now to plant them with only one, usually hawthorn, more rarely blackthorn, occasionally a combination of the two. In particular districts other species were sometimes employed, although they were similarly of little use as fuelwood: the Scots pine, for example, was widely used for new enclosures in the East Anglian Breckland. The significance of the change was succinctly summarised by Arthur Young, writing about Hertfordshire: 'Hedges … cease to be the collieries of a country' (Young 1804, 52).

Changes in farmland species

To a limited extent, the kinds of tree being planted in hedges from the later eighteenth century also changed, reflecting the new species now available to planters and a growing interest on the part of some landowners in ornament rather than production. In 1799 Nicol made a distinction between practical and ornamental planting in hedges, and advocated for the latter the greater use of beech, lime and sycamore. On the Ingilby Estates near Harrogate in Yorkshire, oak and ash remained by far the most common species planted in the hedges, but 'Dutch' elm, pine, black cherry, lime, horse chestnut and hornbeam were also, from the 1780s, sometimes established there, principally as objects of beauty (Dormor 2002, 109). Field evidence – the kinds of tree now growing in field boundaries – indicates that sycamore and beech were widely planted on farmland in the north, especially on large estates such as Bramham, near Leeds, in the course of the nineteenth century. Beech is now a characteristic tree of the northern countryside, in highland as in lowland areas, but had been rare before the later eighteenth century (Figure 6.5). Sycamore may have been more common, in some areas at least, but it also increased in importance through the nineteenth century and is now numerous in many northern districts. In the south there was perhaps less diversification but new species did, nevertheless, appear in the hedges of many farms. In 1827 a sale of timber on a farm in Coltishall in Norfolk – described as 'principally Hedge Row Trees' – included beech, 'abele' [poplar], 'red' poplar and sycamore, although these were considerably outnumbered by

6.5 Beech trees on the Longshaw estate in the south Pennines. Beech has become a familiar feature of many northern and upland landscapes, but before the eighteenth century the species was mainly confined to parts of southern England.

6.6 John Constable's painting of East Bergholt on the Essex–Suffolk border, made in 1813, shows large numbers of Lombardy poplars, looking strangely at home in the English landscape. The tree had been introduced to this country only in the middle of the previous century.

oak and ash (NRO SPE 235, 316 x 1). The 'large quantity of superior Hedgerow Timber Trees' advertised for sale at Bacton in Norfolk in 1868 included, in addition to 100 oak and 100 ash, 150 'sycamore, elm, beech and spruce' (*Norfolk Chronicle* 14 March 1868). Perhaps the most striking, yet transient, addition to English farmland was the Lombardy poplar (*Populus nigra* Italica). This was first introduced in 1758 and became so popular that it appears in some numbers in Constable's paintings of the countryside around the Suffolk/Essex borders, made in the first two decades of the nineteenth century, looking strangely at home there (Figure 6.6). In part, landowners planted it enthusiastically because of its associations with the Italian landscape and its speedy growth, but in part because early advocates asserted the excellence of its timber. In the 1770s a number of individual landowners in Essex, Sussex and Nottinghamshire were awarded Royal Society for the Arts prizes for planting many thousands of them (*Transactions of the Royal Society for the Encouragement of Arts, Manufactures and Commerce* 1782, 3–5). In 1796 Pitt noted that Lombardy poplar had been 'highly extolled some years back' in Staffordshire, but argued that it was a 'mere weed' compared with native timber trees, 'a pole rather beautiful to the eye, but of no promise as a timber-tree' (Pitt 1796, 101). It was not planted only by large landowners, to judge from Constable's paintings, and one commentator in 1808 believed that the tree's reputation had 'suffer'd from its familiarity: because ill-judging people have seen it improperly plac'd', and because it had been widely planted beside the homes of 'the vulgar inhabitants' (private archive).

In spite of such novel additions to the countryside, oak, ash and elm continued to be the principal farmland trees in all areas. In Norfolk, for example, oak and ash made up virtually all the hedgerow timber sold at Wallington in the 1880s and 1890s (NRO MC 111/48, 581 x 5), at Park Farm, Saham in 1906 (NRO RQG 226,489 x 4), at Hethersett in 1905 (NRO RQG 178,489X1) and at Sandringham, Castle Rising, Weasenham and Hargham in 1926 (NRO BL/BG 5/3/22). A survey of 478 trees made on the Heydon estate in 1911, probably a list of farmland timber to be felled, comprised 39 per cent oak, 49 per cent ash and 6 per cent elm; the remaining 6 per cent included maple and wetland species such as poplar, alder and willow, but there were only small quantities of more exotic planting – six sycamore, five beech, three horse chestnut and a single larch (NRO BUL 11/118/1).

Changes in woodland area

Many writers have emphasised the way in which, in the period after *c.*1660, there was a marked upsurge in the planting of new areas of woodland in England, mostly carried out by large landed estates (Prince 1987). Thomas Coke of Holkham in Norfolk, for example, planted around two million trees between 1782 and 1805 (Williamson 1998, 102–3), while Sir Tatton Sykes famously planted some 1,000 acres (400 hectares) with trees on his estates in the Yorkshire Wolds, a memorial tablet in West Heslerton church proudly proclaiming: 'Whoever now traverses the Wolds of Yorkshire and contrasts their present appearance with what they were cannot but extol the name of Sykes' (Ward 1967, 13). An enthusiasm for planting had begun in the later seventeenth and early eighteenth centuries, fuelled by

6.7 Harewood in Yorkshire, showing the extensive eighteenth-century plantations within the park. Eighteenth-century estate planting was often concentrated in, and in the immediate vicinity of, landscape parks laid out around great mansions.

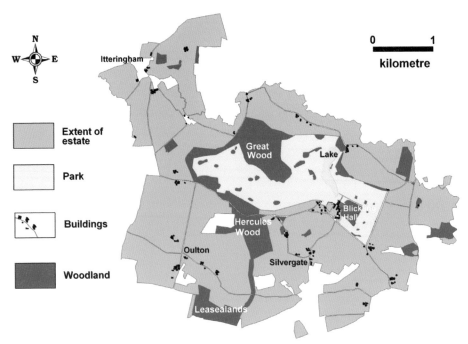

6.8 Woods and plantations at Blickling in Norfolk, as they were in 1840. 'Great Wood' is in part an area of ancient, semi-natural woodland. The others were planted in the course of the eighteenth and nineteenth centuries.

a rash of texts on forestry, including John Evelyn's *Sylva, or a Discourse on Forest Trees* of 1664, Moses Cook's *The Manner of Raising, Ordering and Improving Forest Trees* of 1676 and Stephen Switzer's *Ichnographica Rustica* of 1718. By the second half of the eighteenth century the Society of Arts was offering annual medals in recognition of particularly extensive afforestation projects. Planting was carried out to make money, but it had other motivations. It was seen as a patriotic duty, for there was widespread concern that there was a general timber shortage, which had implications for the nation's naval power, as writers such as Phillip Miller (1731), James Wheeler (1747), Edmund Wade (1755) and William Hanbury (1758) all warned. In a more general sense the planting of trees demonstrated confidence in the new political dispensation brought about by the Restoration of the Monarchy and the Glorious Revolution of 1688 (Daniels 1988). Tree planting likewise expressed confidence in the continuity of ownership on the part of local dynasties, and was in itself a sign of status. As already noted, large-scale tree planting was intimately associated with large estates because only those owning extensive properties could afford to put hundreds of acres out of agricultural production, foregoing immediate for medium- or long-term financial benefit. But planting was also carried out simply to beautify estates, especially as the fashion for 'landscape parks' took hold from the 1750s and 1760s, under the influence of Lancelot Brown and his contemporaries (Figure 6.7) (Brown and Williamson 2016). Such parks proliferated steadily in the second half of the eighteenth century, and woodland clumps and belts were a key feature of their design. New plantations were also established to provide cover for game, especially as the pheasant, a woodland bird, became the main quarry of the hunter in the second half of the eighteenth century. Many of these likewise took the form of clumps or belts – the pheasant being, primarily, a bird of the woodland edge (Hill and Robertson 1988). Much of this new planting, as we shall see, involved new species introduced from abroad. In 1550 around 36 introduced hardy and woody species existed in England; this had increased by 1600 to 103, by 1700 to 239, by 1800 to 733 and by 1911 to no fewer than 1,911. Of course, most of these were ornamentals, for gardens and parks, but some were primarily employed in commercial plantations (Watkins 2014, 67–8).

So far as the evidence goes, it was only really from the 1780s that aristocratic planting schemes extended much beyond the bounds of parks and the immediate vicinity of great mansions: indeed, on many estates the bulk of woodland remained concentrated there (Figure 6.8). Nathaniel Kent observed in 1796 that while 'gentlemen of fortune' in Norfolk had carried out much tree-planting 'in their parks and grounds', the planting of 'pits, angles, and great screens upon the distant parts of their estates, which I conceive to be the greatest object of improvement, has been but little attended to' (Kent 1796, 87). A little earlier William Marshall noted that in Gloucestershire new planting was concentrated in parks and had 'all been done with a view to ornament, merely. The propagation of wood, for the use of estates' had scarcely begun, even though the shortage of wood was so serious in the locality that firewood was brought a distance of 10 miles and coal as much as 30 (Marshall 1789, 26). Even in 1810 it was thought that in Durham new plantations had been made 'especially in the vicinity of gentleman's seats' (Bailey 1810, 191). It was the large-scale parliamentary enclosures of heaths and moors, and to an extent of open fields and sheepwalks in areas of poor, thin soils, that occurred in the decades either side of 1800 that provided particular opportunities for afforestation, for plantations made good economic use of such marginal land. The Fairbanks family, surveyors in the Sheffield area, claimed in the early nineteenth century that 'proprietors in many instances derive more profit from the sale of wood, than could be obtained from the rent of land if cultivated' (Shf FB/CP/25/1–2). Such concern for private profit shaded easily into wider patriotic considerations and an enthusiasm for 'improvement'. One Yorkshire commentator in 1823 thought that it was 'to be regretted that so much barren land should be suffered to remain unplanted … and ought not to lie useless to the Community when the soil could be advantageously employed' (*Hatfield Mercury* 16 August 1823). There was also a degree of social emulation involved, with landowners vying with neighbours to demonstrate their enthusiasm for this, as for other branches of 'improvement'. As Strickland, commenting on the afforestation schemes carried out by the owners of the Bolton estate in the West Riding of Yorkshire, noted: 'others, encouraged by their success, and the increasing demand and value of timber, continue to plant, and some even with the sole object of profit' (Crowther 1992, 62).

For the most part, in fact, afforestation projects had mixed and complex motives, and any distinction between 'aesthetic' and 'economic' forestry can be over-drawn. The regular fellings made as plantations were thinned, as described below, ensured that even the clumps and belts in parks were regarded as a source of income as well as objects of beauty, and all estate planting represented, to some extent, money in the bank. The great belt planted around the perimeter of the park at West Tofts, in south-west Norfolk, was put up for sale along with the rest of the estate in 1788, and the sales particulars described at some length the 'most agreeable ride' which ran along its centre. Yet they also noted how:

> The number of trees that will remain in the Plantations, after they are thinned so as to leave them at a proper distance, to facilitate their Growth, will be about Six Hundred Thousand: which in the Course of a few years, will at least be worth a shilling a Tree, and consequently amount to Thirty Thousand Pounds. (NRO MC 77/1/521X7)

The impression given in many published sources is that the 'great replanting' of the eighteenth and nineteenth centuries led to a very significant expansion in the area of woodland in England (Hadfield 1967, 79–174). But in reality the situation was more complicated. The extent to which new woods were established in different districts, especially in the period after *c.*1780, varied according to a range of factors. These included, crucially, the extent and quality of land newly enclosed, especially from heaths and moors; the extent to which land in any area was owned by large estates, for in many districts significant number of small proprietors – freehold farmers or minor gentry – continued to exist; and the amount of woodland already present in the landscape, for where woods were already abundant, on the whole, the scale of planting was less. More importantly, new planting was also balanced by the loss of existing woods, to varying degrees in different districts, in this age of agricultural 'improvement'. Methods of improving soil quality through marling and, in particular, novel forms of land drainage now made possible the profitable cultivation of many of the sites occupied by woods at a time of steadily rising agricultural prices (Wade Martins and Williamson 1999, 61–7). In addition, by the late eighteenth century grazed woods – wood-pastures – were becoming redundant in economic terms and were regarded as particularly old-fashioned features of the landscape by the social elite. Surviving tracts – especially in forest districts – were systematically destroyed following enclosure, converted to agriculture or (more rarely) to plantations, as critics had long urged. The interaction of these varied factors ensured that in reality the area of woodland in different districts and regions of England increased to very varying degrees – and in some cases actually declined – although our sources allow us to estimate the changes only in very broad terms.

In Norfolk, for example, the area of enclosed woodland shown on William Faden's reasonably accurate county map of 1797 amounted to some 13,417 hectares, or 2.4 per cent of the county's surface area. At most, surviving wooded commons accounted for a further 670 hectares, bringing the total area of woodland in the county in the late eighteenth century to around 14,000 hectares, or around 2.5 per cent of its land area. Although these figures need to be treated with caution there is little doubt that by 1895 there had been a substantial increase in woodland cover, to around 4 per cent (Board of Agriculture Returns, 1896, 36). Some of this comprised aesthetic and amenity planting, such as belts and clumps in landscape parks, but much represented large-scale afforestation of marginal land, mainly former heathland enclosed in the decades either side of 1800. In this county, such additions to the woodland area were only to a limited extent counter-balanced by the grubbing out of wood-pastures or long-established coppice woods, although there were examples of both. Catfield Wood in the east of the county was thus destroyed altogether in the early nineteenth century; Hedenham Wood lost around a third of its area, Horningtoft Great Wood more than two-thirds and Rawhall Wood in Beetley over a quarter in the first three decades of the nineteenth century; while between c.1840 and 1880 Ashwellthorpe Wood, Banyards Wood in Bunwell, Billingford Wood, Horningtoft Great and Horningtoft Little Woods, Old Pollard Wood in Holt, North Elmham Great Wood and Shropham Grove were all significantly truncated (Barnes and Williamson 2015, 113). The overall effect of new planting in the county, amplified by the grubbing-out of some woods, was to even out the distribution of woodland between different districts, as woods were lost from the claylands but proliferated on lighter soils. Only in the drained wetlands of the Fens in the far west and the Broads in the east did woods and plantations remain rare.

Other regions of England in which large areas of marginal land were enclosed in the period after c.1780 saw similar increases in woodland area. Yorkshire was, by the later eighteenth century, rather sparsely wooded, although the county map surveyed by Jefferys and published in 1771 (Jefferys 1771) is too schematic in its treatment of woodland to provide a reliable impression of its extent, and no draft Ordnance Survey drawings from the early nineteenth century – a useful source in most parts of the country – exist for the north of England. Local maps, however, dating from the decades either side of 1800, show few woods and contemporaries were united in their description of the county as singularly lacking in timber. In the East Riding Leatham in 1784 believed that the Wolds were 'not generally adapted to the growth of wood, and very little is to be found', while in the lowlands immediately to the north, beside the rivers Hertford and Derwent, there was 'but little wood' (Crowther 1992, 35). Even in 1812 Henry Strickland was able to claim that the trees growing in the woods and hedgerows on one estate in Escricke were together worth more than all the full-grown timber in the East Riding put together (Crowther 1992, 61), while the district of Holderness lying to the east of the Wolds 'never contained any woods'. Only in the strip of land between the rivers Ouse and Humber and the Wolds, and to the south of York, could any significant areas of woodland be found (Crowther 1992, 34–6). In the case of the North Riding, Tuke in 1800 estimated that there were only 25,500 acres (10,319 hectares) of woodland – less than 2 per cent of the total land area – and Jefferys' map of 1771 suggests that this was mainly concentrated, as it had long been, on the slopes of the deeply incised valleys draining from the North York Moors to the Vale of Pickering and in the Hambleton and Howardian Hills. There was more woodland in the West Riding, clustered in particular on the flanks of the Pennines, although even here, to judge from contemporary maps, only marginally more than in the other Ridings, and certainly occupying no more than c.3 per cent of the land area.

By 1895 the total area under woods and plantations in the East Riding had reached 17,181 acres (6,953 hectares), some 2.3 per cent of the surface area (Board of Agriculture Returns, 1896, 36). The vast majority of this must have been planted during the previous century or so. The North Riding now boasted 52,816 acres (21,374 hectares), more than double Tuke's estimate. In the West Riding there was a similar amount of woodland – 69,592 acres (28,163 hectares), or 3.9 per cent of the land area (Board of Agriculture Returns, 1896, 36) – much of which had evidently been planted by large estates following the enclosure of moorland in the early nineteenth century (Dormor 2002, 232). On the Bolton estate near Wensleydale, for example, Lady Bolton and Sir John Orde established some 30 new plantations, containing around 55,000 trees, in the early years of the nineteenth century (Dormor 2002, 232). On balance, it seems likely – allowing for the fact that woods were lost as well as gained – that the area of woodland in Yorkshire as a whole increased by more than a third in the century after c.1780, from around 2.5 to 3.5 per cent of the land area.

While sparsely wooded regions thus saw significant rises in woodland area, those already well-endowed with woodland saw more stability, or even a slight decline. In Hertfordshire, for example, a comparison of the draft Ordnance Survey drawings from the early nineteenth century with the first edition 6-inch map from the 1880s shows many instances of woodland reduction, mainly involving small areas (especially narrow 'hedge rows') but including some larger examples. These were perhaps most marked on the boulder-clay soils in the east of the county, where (for example) no fewer than 11 small and medium-sized woods, with a combined area of 27 hectares, disappeared in the adjacent parishes of Much and Little Hadham between the early nineteenth century and the 1880s (HALS DE/Cn/P4; HALS DE/X713/P1). The wood-pasture commons in the south of the county, and many of those in the Chilterns, were also destroyed following enclosure around 1800. Yet, at the same time, there were many piecemeal additions of woods and plantations, especially within and on the margins of the landscape parks that were particularly numerous in this county, on account of its proximity to the metropolis. Overall, the area of woodland (including wood-pasture) probably declined in the century after 1780, from around 11,000 hectares, or c.7.5 per cent, of the county's land area to the 10,338 hectares, or 6.3 per cent, recorded in 1895 (Board of Agriculture Returns, 1896, 36); but, given the somewhat schematic character of the earlier sources, it is perhaps safer to say that there was little significant change.

    In some counties, in contrast, there were very significant reductions in woodland area. In Northamptonshire, for example, a systematic reconstruction, based on a wide range of local maps, suggests that there were around 13,500 hectares of coppiced woodland in c.1750, c.5.2 per cent of the total land area, together with perhaps a further 4,000 hectares of wood-pasture, amounting to another c.1.6 per cent.[2] The vast majority of both continued, as in earlier centuries, to be concentrated in the three royal forests of Salcey, Whittlewood and Rockingham, all occupying boulder-clay uplands between the principal river valleys. These woods and wood-pastures survived reasonably intact until the enclosure of the forests, by a series of parliamentary acts, in the nineteenth century. Brigstock Bailiwick in Rockingham Forest was enclosed in 1805 under an act passed in 1795 and Cliffe Bailiwick in 1806 under an act of 1796, but Salcey Forest was enclosed only in 1826, the rest of Rockingham in 1837 and Whittlewood in two stages, with awards in 1826 and 1856 (Pettit 1968; Williamson *et al.* 2013, 144–8). The acts in all cases emphasised the 'injurious' effects of common rights on the value of the underwood in the enclosed coppices and on timber, and the Salcey award specifically mentions that the allotments to the Crown and the Duke of Grafton were to be selected on the basis that they would be the best areas for establishing plantations (NHRO X1693 Bundle 21; G3909). But it was also envisaged that many of the allotted areas could be cleared and cultivated, and the main result of enclosure was indeed a rapid diminution in woodland area. The wood-pasture plains were destroyed immediately after enclosure, and around 5,680 hectares of coppice had been converted to farmland (or occasionally to parkland) by the time the Ordnance Survey 25-inch maps were surveyed in the 1880s. To some extent such losses were offset by the establishment of new woods elsewhere in

2   This was carried out as part of an AHRC-funded investigation of changing patterns of land use in Northamptonshire. See Williamson *et al.* 2013 and http://archaeologydataservice.ac.uk/archives/view/midlandgis_ahrc_2010/.

the county by major landowners. But, on the other hand, some of the relatively few long-established coppiced woods lying away from the forests also disappeared in this period, including several in the area around Stow Nine Churches and Farthingstone, in the west of the county. Overall, the area of woodland in Northamptonshire, including wood-pasture, probably shrank from around 17,500 hectares in the later eighteenth century to around 11, 338 in 1895 (Board of Agriculture Returns, 1896, 36): that is, from around 7 per cent to around 4.5 per cent of the land area.

The very varying experience of these four counties shows that we need to be cautious in accepting uncritically the idea that there was a significant increase in woodland area in the course of the later eighteenth and the nineteenth centuries as a consequence of aristocratic afforestation. Even in 1895 only around 5 per cent of the land area of England was occupied by woodland; in c.1780 the figure may have been around 4 per cent, including wood-pastures, suggesting a rise at most of around 20 per cent in the century of the 'great replanting' – a rather less dramatic increase than is sometimes suggested or implied (Hadfield 1967; Daniels 1988). More important, perhaps, were changes in the

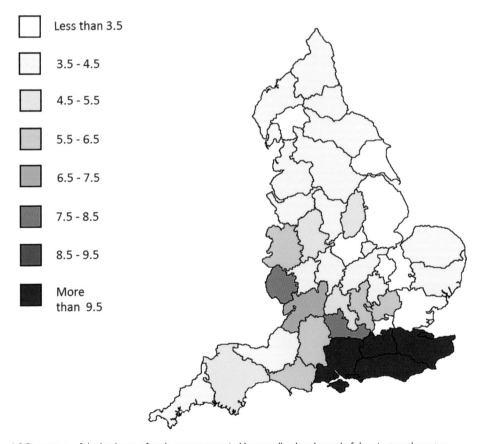

Less than 3.5

3.5 - 4.5

4.5 - 5.5

5.5 - 6.5

6.5 - 7.5

7.5 - 8.5

8.5 - 9.5

More than 9.5

6.9 Percentage of the land area of each county occupied by woodland at the end of the nineteenth century (based on the Board of Agriculture figures for 1895).

distribution of woodland that resulted from loss and addition. Increases were greatest in areas where formerly there had been few woods, and especially in districts of more marginal land: on thin chalk soils following the enclosure of open fields and sheepwalks; on poor sandy soils after the enclosure of heaths; and on upland moors. As Gooch noted of Cambridgeshire, a mainly 'champion' county, in 1811: 'Woods are not extensive in this county. Many plantations have been made where enclosures have taken place; they have not only considerably improved the face of the country, but promise to prove a source of profit … on poor, light soils' (Gooch 1811, 197).

The greatest losses of woodland, in contrast, were on heavy but fertile clay soils, and especially on those formed in boulder clay in the Midlands and East Anglia – as in Northamptonshire, where the three royal forests all occupied clay uplands, and also as already described in Norfolk and Hertfordshire. All this was part of the general environmental homogenisation that occurred in England during the period of the agricultural revolution, as what we now think of as the 'typical' English countryside of enclosed fields interspersed with areas of enclosed woodland replaced more dramatically diversified landscapes, featuring extensive tracts of wetland, heath and wood-pasture. This said, at the end of the nineteenth century woodland was by no means evenly distributed across England. Woodland was concentrated in a band running from the south of England into the west Midlands, largely within old-enclosed 'woodland' districts. In the north woodland still remained relatively sparse, as it did in the eastern counties – in the main arable-farming districts extending from the East Riding of Yorkshire to Essex – a distribution cutting resolutely across any ancient distinction between 'woodland' and 'champion' districts (Figure 6.9).

Plantation forestry

Most of the new woods established in this period were *plantations*, containing only timber trees and lacking a coppiced understorey. In the majority of cases deciduous species – old-established mainstays of forestry such as oak and elm, but also sweet chestnut, beech and sycamore – were mixed with a rather larger number of 'nurses', most of which were conifers, especially Norway spruce and European larch (introduced in the sixteenth and seventeenth centuries respectively) and Scots pine. In the words of Nathaniel Kent, plantations in Norfolk consisted of 'Great bodies of firs, intermixed with a lesser number of forest trees' (Kent 1796, 87). On the Ingilby estates in north Yorkshire the entries in Sir John Ingilby's planting book for 1781 record the planting of a 'great quantity' of oak, ash, beech, birch and holly, together with 1,000 sycamore, as well as 600 larch, 700 spruce and as many as 3,500 Scots pines (there were also smaller quantities of alder, willow, beech, Dutch elm, Weymouth pine, black cherry, lime, horse chestnut, hornbeam and ash, probably planted for aesthetic reasons). In 1783 2,290 oaks were planted on the estate, but these were accompanied by 5,470 Scots pine and 3,160 larch, together with 1,400 silver fir, 150 horse chestnut, 206 Weymouth pine and a mere 70 ash (Dormor 2002, 106–7). The Fairbanks family were advised in the early nineteenth century that on moorland in south Yorkshire 'the larch will probably pay best in both thinnings & timber', although 'experienced planters aware of the great uncertainty in regard to any criterion for practice will always plant a variety of sorts some of which will scarcely fail to make timber in any situation'. It was suggested that they plant '½ larch & the other ½ scotch fir with spruce fir in moister parts, for the noble tree the spruce will not succeed in dry ground; oaks, elms, beeches, birches, in equal proportions'. The plantation would therefore consist of '1550 larches, 300 of each of the other makes abt [*sic*] 3000 plants on an acre. Every other row as wholly larch, the intervening rows mixed' (Shf FB/CP/25/1–2). In most upland areas larch seems to have been the favoured conifer in mixed plantings. Culley in 1797 described how in Northumberland 'the larch rises proudly pre-eminent above the rest, and in almost every situation far outstrips the various specimens of firs and pines' (Bailey and Culley 1794).

Some of the new plantations appear to have comprised conifers alone, most commonly Scots pine, spruce and larch, alone or in combination, but sometimes other species, as at Westwick in Norfolk, where those established to the north-west of the park by John Berney Petre were reported, in the 1790s, to consist entirely of 'pinasters [Maritime pine or *Pinus pinaster*], except in the valley where other trees grow' (Grigor 1841, 154). An 1806 valuation of the timber growing at Mundford, in the Norfolk Breckland, described one plantation – 'Long Plantation' – that consisted, in the manner just described, of a mixture of larch, pine, fir, elm, oak and birch. But two others were planted with conifers alone: 'Square Plantation' contained 9,827 larch, spruce, Scots pine and Weymouth pine, while 'Round Plantation' consisted of 4,360 larch, spruce and Scots pine (NRO MS 13751, 40E3). In upland districts, too, purely conifer plantations were often established. Many of those planted by George Browne around Troutbeck in Cumberland in 1774, for example, consisted entirely of 'firs', larch and spruce (Parsons 1997, 91).

Whatever the precise composition of new plantations, the trees were usually planted more closely than would be usual today. It was reported that on the Longshaw Estate in the Peak District they were planted in 'rows 4ft from row to row & 3ft 6 inches from plant to plant. This is much better for cutting out the thinnings, which after all is the main thing in this business & they make clean timber with less pruning' (Shf FB/CP/25/1–2). Tuke in 1800 similarly described how Yorkshire plantations were composed of 'forest-trees of two years old, and planted at about four feet distance' (Tuke 1800, 181). But planting distances of three feet, or even less, are also reported (Bailey 1810, 199). Such dense planting was necessary because it was difficult to deal with the weeds that competed with the young trees, and also to some extent because significant losses were anticipated from drought and the depredations of rabbits and other animals. Indeed, the journals of eighteenth- and nineteenth-century landowners often describe their forestry activities as a heroic battle against the odds of nature, William Windham of Felbrigg in north Norfolk, for example, bemoaning in 1826 how 'The hares and rabbits have destroyed the plantations on Harrisons Brake entirely, not withstanding the great cost of the fencing' (NRO WKC 7/134). Secure fencing was, indeed, a *sine qua non* of successful forestry, but in upland areas especially drainage was also an important consideration.

Having been planted densely, the trees were then progressively thinned, beginning at around five years and continuing at frequent intervals thereafter. The fact that the extracted material is often referred to in estate accounts as 'poles' signifies, clearly enough, that it was used in a similar way to the produce of coppices (Williamson 1998, 183–6). Nathaniel Kent described the great plantation belt around the park at Holkham as comprising 'four hundred and eighty acres of different kinds of plant, two thirds of which are meant to be thinned and cut down for *underwood,* so as to leave the oak, Spanish chestnut, and beech, only as timber' (Kent 1796, 90: our italics). In mining areas the larger thinnings were often consumed by the local mines, especially as pit props. In many cases the final timber crop was itself only partially removed, if at all, leaving the plantations to provide shelter or cover for game, or to beautify the countryside. Careful attention needed to be paid to thinning at the appropriate time, otherwise the trees grew too cramped, and in mixed plantations the deciduous species might be over-topped by the faster-growing conifers. Marshall in 1787 described the poor state of a plantation established 25 years earlier on the Gunton estate in north Norfolk:

> The *Scotch fir* has outgrown any other species; and the plants, though few, have become a burden to the grove. The wood being of quick growth, the plants have not only out-topped the rest, but have, in general, had time enough to furnish themselves with boughs on every side … . The *larches*, too, where they stand free from the Scotch firs, are of considerable size. (Marshall 1787, II, 95)

The new forestry involved the widespread planting of coniferous species introduced from abroad. Some of these, such as the European larch and the Norway spruce, had been present in the country for over a century, originally being treated as ornamentals, but others – such as the Sitka spruce, introduced in 1831 – were specifically brought into the country with commercial forestry in mind. Perhaps equally important, however, is the fact that the eighteenth and nineteenth centuries saw the spread of indigenous or long-naturalised trees into new parts of the country. Beech, for example, appears to have long been common in the south of England, especially (as we have seen) in the Chilterns, the Cotswolds and the area around London. But it was rare in the Midlands and East Anglia, and in the north virtually unknown. From the later eighteenth century it was planted extensively in all these areas. Sweet chestnut, naturalised at an early date in the south of the country, only now spread more widely, becoming a popular tree for forestry. Scots pine similarly extended its range from Scotland southwards across all parts of England from the seventeenth century. It is not entirely clear how far the enthusiasm for these species was grounded in economics and how far it was dictated by aesthetic concerns and fashion, but beech, as we have seen, was a useful wood, widely used for boards and furniture making, and, while it appears to have fetched prices around half those for oak, it was a faster-growing tree. This said, much appears to have been extracted and sold at pole stage, and in many contexts the final 'crop' was perpetuated as an adornment of the landscape. Sweet chestnut was likewise a useful wood, especially for fencing, but again its aesthetic qualities may have been a major reason for the enthusiasm with which it was planted by landowners.

The impact of all these new species on estate planting should not be exaggerated. A survey made in 1816 of the timber in the plantations outside the park on the Houghton estate in Norfolk listed 13,242 trees comprised of no fewer than 18 different varieties (alder, ash, beech, birch, cherry, cedar of Lebanon, elm, 'firs', hornbeam, horse chestnut, larch, lime, oak, Scots pine, spruce, sweet chestnut, sycamore and white poplar) (Houghton Hall archives) – but oak, ash and elm made up two-thirds of the total. After oak (50 per cent) and ash (13 per cent), Scots pine (10 per cent) was the next most important tree; no other species accounted for more than 5 per cent, and all the conifers combined made up only 15 per cent of the total listed. In Yorkshire, similarly, oak continued to be the dominant tree recorded in nineteenth-century surveys, followed by ash. Out of c.73,600 trees (including poles, 'cyphers' and the like) whose species is described in nineteenth-century valuations, 48 per cent were oak and 21 per cent ash. Larch made up the third largest category. Even in the early twentieth century oak, ash and elm continued to be the most important trees in the woods and plantations on most estates, at least in the south and Midlands. At Heydon in north Norfolk in 1921, for example, oak and ash made up 73 per cent of the 728 trees sold from the estate woods: the next most numerous species was beech (6 per cent), followed by sycamore (2.6 per cent) (NRO BUL 4/118, 615x9). This said, on the poorer, more acid soils in particular, conifers and species such as beech, sycamore and sweet chestnut were locally numerous. Also in Norfolk, a list of trees uprooted in a storm in 1916 on the Bagge family's properties on the poor, sandy soils around Bawsey, to the east of King's Lynn, listed 12 oak and 9 elm, together with 7 beech, 6 poplar, a lime and a sweet chestnut, all outnumbered by 77 conifers of various kinds, in addition to the 'many Pine, Spruce and Larch' noted as lost in one of the plantations (NRO BL/BG5 /3/18).

The fate of coppiced woods

It might be expected that the fashionable interest in the 'new forestry' would have led to the general neglect of coppiced woodland, and this, indeed, is the impression conveyed by many modern texts. Even contemporaries sometimes effectively ignored traditional forms of management. Nathaniel Kent's book on the agriculture of Norfolk, published in 1796, for example, devotes 319 lines to plantations (with a further appendix devoted to those composed of sweet chestnut) but just 19 to woodland in the traditional sense (Kent 1796). Moreover, as already noted, many areas of existing woodland, mainly of ancient, coppiced type, were lost or truncated in the course of the eighteenth and nineteenth centuries. But it is striking that most of the authors of the *General Views* of the agriculture of various counties, published in the decades around 1800, devote many pages to coppices and, far from suggesting that they were an old-fashioned form of land use, generally vaunt the profits that could be derived from them. The tithe files from the 1830s similarly suggest that most woods were coppices, and still actively managed as such.

It was mainly in northern areas, where labour costs were high and the use of coal, along with industrial substitution, reduced demand for underwood, that coppicing seems to have declined in importance during the nineteenth century. By the end of the eighteenth century lead ore hearths were already being replaced by furnaces fired by coal, and coke had largely replaced charcoal in iron smelting. While the market for whitecoal declined, that for charcoal survived to an extent because it was employed in the manufacture of blister steel (Jones 2012, 65). However, demand for coppice wood underwent a further contraction from the 1860s owing to a fall in the price of bark, as new methods of industrial tanning, utilising chromium salts, were adopted. From the middle of the nineteenth century regular coppicing was thus replaced with long-term management for timber, and large swathes of woodland were converted to high forest. They were either replanted or oak coppices – already usually cut on a long rotation – were 'singled' or 'stored', so that only one of the poles on each stool was left to grow to maturity (Jones 2012, 65–7). 'By the end of the nineteenth century, in many northern districts, coppicing, and its related trades and industries, had all but disappeared' (Jones 2005, 55).

Yet in many districts, especially in the south and the Midlands, coppicing continued more strongly through the nineteenth century, in spite of the greater use of coal – a clear enough indication that woods were not solely sources of firewood but also suppliers of wood for specialised uses. While the tithe files from the 1830s and 1840s certainly describe in a number of places how coppices were suffering owing to the density of the timber trees, suggesting that the production of wood was taking second place to that of timber, for the most part local sources suggest that active and profitable management of the underwood continued. While the arrival of a canal or, in particular, a railway in a district lowered the value of fuelwood, it might raise that of poles destined for other, more specialised uses. One landowner in Herefordshire reported in 1852 how 'a railway is now being formed through the district which will be a means of bringing coals into the neighbourhood at a reduction of at least 8/- per ton'; he acknowledged that this would reduce the price of firewood, but believed that 'it may also cause some of the produce of the woods to be taken to distant places – to be used for lath hoops, casks, handles for working tools, and other things' (HAR A63/111/56/12). A poster drawn up in the following year to advertise the sale of a coppice wood in the same county emphasised that it lay 'within a short distance of the proposed line of Railway from Shrewsbury to Hereford' (HAR F76/111/50).

Specialised uses for coppice poles continued to be stimulated by economic or industrial developments. Collins has suggested that as many as 1,500,000,000 bobbins may have been required by British textile mills each year by the middle of the nineteenth century, at least half of which came from coppices. In the south of England the continued growth in hop production ensured that by the middle of the century the income from coppiced woods might reach £3 or more per acre, per annum, roughly twice that which might be derived from agricultural land of similar quality (Collins 1989, 488–9). But even without the stimulus provided by such specialised demands, the value of coppice often held up well. Timber accounts from the Stow Bardolph estate in west Norfolk, for example, show that the estate woods were still being regularly coppiced in the 1860s: there are records of sales of pea sticks, stakes, poles, faggots and, above all, hurdles from the two main coppiced woods on the estate, Spring Wood and Toombers, the latter fetching £3 16s a dozen (NRO HARE 5501, 223 x 1). Of particular interest is the description provided in 1851 by Henry Wood, the agent for the Merton estate, of the management of Wayland Wood in Watton in the same county. He related how the larger poles were used to make hurdles, fencing or bins for storing hay or straw on the home farm; how other material was cut as splints, about six feet in length, that were used for building repairs; and how the smaller material went for thatching broaches and sways and for pea sticks for gardens. The residue was used as fuel, the smaller brush faggots being sold to bakers and cottagers for 'oven wood' and the off-cuts going for 'cottage firing' (NRO WLS XVIII/7/1). On other estates, throughout the nineteenth century, poles continued to be allocated to tenants for farm repairs, gates and fencing (as, for example, on the Evans-Lombe estate in mid-Norfolk: NRO HNR 112/19). As late as 1871 the fourth and fully revised edition of James Brown's *The Forester* included sections not only on the management of existing coppices but also on the establishment of new ones. Coppices were 'very much cultivated' in England, forming 'a large proportion of crop given over to wood' (Brown 1871, 610). The money that might be made by selling the offcuts and 'small rubbish' as fuel was emphasised, at least 'in inland counties, where coal is high-priced'.

Indeed, in the course of the nineteenth century attempts were often made to improve long-established coppiced woods by, in particular, installing networks of land drains within them. These usually took the form of ditches around 0.4 metres deep and between 1 and 2 metres in width that were often laid out in dense and regular networks (Figure 6.10). Such schemes were implemented in the largely erroneous belief that they would improve productivity in the same manner as the underdrains that were by now widely employed on agricultural land. Brown, for example, urged that 'a properly executed system of drainage will result in heavier and more profitable tree-crops being yielded' (Brown 1847, 531). The fact that the drains were usually dug ruler-straight – no easy task in a wood containing numerous coppice stools and sporadic timber trees with extensive root systems – clearly betrays the hand of faddish, 'scientific' improvement.

On large estates an additional reason for the continuation of traditional management in coppiced woods was that the underwood, regularly cut, provided good game cover. Whereas plantations, as they were thinned and matured, needed to be underplanted with rhododendron or other shrubs, the larger coppice woods, felled on rotation, always had a significant part of their area suitable for the roosting birds. Indeed, the owner of Wayland

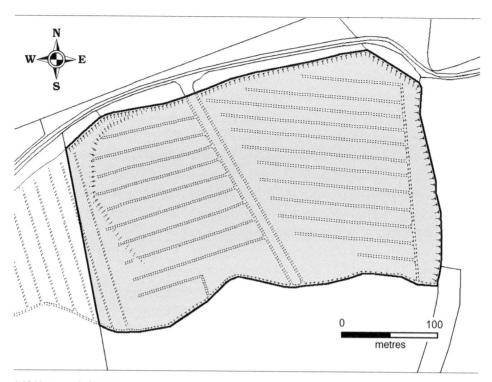

6.10 Horningtoft Great Wood, Norfolk. The wood was considerably reduced in area, in stages, in the middle of the nineteenth century, like many ancient woods occupying fertile clay soils. The dense network of drains was installed between two of these phases of removal, and thus extend into the pasture field, formerly part of the wood, to the west. The drains have the typical ruler-straight character associated with nineteenth-century 'improvement'. They would have been difficult to establish among the scattered coppice stools and timber trees.

Wood in Norfolk, Lord Walsingham, described in 1893 in some detail how the wood was beaten, commenting that

> in about four years after felling, the condition of the undergrowth will be most suitable for this purpose … . In some portions of a large wood, the undergrowth will be from five to ten years old, and much of this will be difficult to beat through and more difficult to shoot in where paths or rides are cut. (Walsingham and Payne-Gallwey 1893, 220)

Miche suggested in 1888 that 'the principal reason for maintaining old woods near the mansion' was that there was 'never at any one time … so much wood cut down as to cause serious blanks or openings' (Miche 1888, 105). In part because of their increasing importance as coverts, many woods were taken back into direct management by their owners from the later decades of the eighteenth century. Where they were not there was obvious potential for conflict, as at Helhoughton in Norfolk in 1813, where it was alleged that 'William Withers … caused the whole of the underwood which was growing … to be cut … which was very prejudicial to the game' (NRO BUL 11/436). In addition, throughout the eighteenth and nineteenth centuries established coppiced woods were – like farmland trees (Rackham 2004) – often incorporated into the designed landscapes laid out around great houses, forming part of the perimeter belt or utilised as internal clumps. Their use in these ways again tended to encourage the continuation of regular coppicing in order to provide a solid mass of vegetation to close a view or to provide a sharp and continuous outline.

Such was the enthusiasm for traditional forms of woodland management that, well into the nineteenth century, completely new areas of coppice-with-standards were sometimes planted by large landowners. Many were established in long-enclosed areas, where 'ancient woodland indicator' herbs, such as dogs mercury (*Mercurialis perennis*), bluebells (*Hyacinthoides non-scripta*) or wood anemone (*Anemone nemorosa*), were often present in adjacent hedgerows. Such plants spread with remarkable rapidity into the new woods, encouraged by regular coppicing and the absence of grazing. As a result, some woods planted in this period can be hard to distinguish from 'genuinely' ancient woods (Stone and Williamson 2013; Barnes and Williamson 2015, 122–33). Lopham Grove in south Norfolk comprises coppice of hazel, ash, maple and hornbeam under standards of oak (*Quercus robur*), with some ash and the occasional sweet chestnut. The ground flora contains no fewer than six 'ancient woodland indicators': dog's mercury, wood spurge (*Euphorbia amygdaloides*), wood sedge (*Carex sylvatica*), primrose (*Primula vulgaris*), bluebells (*Hyacinthoides non-scripta*) and early purple orchid (*Orchis mascula*). Yet a map of 1720 shows the entire site as completely unwooded (Arundel Castle Archives 5H3 18). Primrose Grove and Old Grove in Gillingham in the same county – conjoined areas of ancient woodland that were planted between 1840 and the 1880s – comprise ash, hazel and hornbeam coppice and have a ground flora that includes dog's mercury, primrose, dog-violet (*Viola reichenbachiana*), hard shield-fern (*Polystichum aculeatum*) and wood speedwell (*Veronica montana*) (Barnes and Williamson 2015, 218–19). The neat distinction usually made between 'ancient' woodland – planted before 1600, formerly coppiced and with an extensive array of 'ancient woodland indicators' – and woods newly planted in the eighteenth or even the nineteenth century is exaggerated in many modern texts.

Age of felling

Perhaps the greatest single change in the management of trees in England in the course of the nineteenth century, other than the virtual disappearance of pollarding, was a marked rise in the age at which timber – in woods and especially on farmland – was felled. As we have seen, in the period before the mid-nineteenth century timber trees were generally felled before they exceeded 60 years of age, at which point most will have attained – very roughly – a volume of around 50 cubic feet. Because many trees were taken down when they were much younger than this, average estimated volumes of trees growing in or extracted from woods or hedges were commonly less than 15 cubic feet for ash and 20 for oak. By the end of the century, in contrast, the average age of felling, and thus the average age of trees more generally in the landscape, had risen markedly. This development is obscured to some extent by management practices in plantations, where average age continued to be depressed to some extent by the small size of 'timber' trees extracted in the later stages of thinning. But in woods of coppice-with-standards type, and on farmland, the change is very clear. To give some examples from Norfolk, the average volume of oak and ash trees felled on the Anmer estate in 1881 was 47 cubic feet (NRO MC 40/95), while that of the 184 trees taken down in Oak Wood, Little Cressingham, in 1895 was just under 49.5 cubic feet (NRO HILS 3/250). All the oaks felled on the Watlington estate in the 1880s and 1890s contained between 40 and 50 cubic feet (MC 111/48, 581X5), and by the 1920s at Heydon the average volume of felled oaks was around 50 cubic feet (NRO BUL 11/118/11; BUL 4/118). Many more timber trees were thus being allowed to grow up to, and beyond, maturity.

To some extent this change may reflect new aesthetic preferences on the part of large landowners, for a landscape dominated by full-grown standard trees, no longer managed primarily with economics in mind – attitudes which began to develop from the later eighteenth century, with the rise of 'Picturesque' aesthetics under the influence of William Gilpin and others, and grew in importance through the first half of the nineteenth century (Watkins 2014, 109–13). But for the most part it probably indicates that imports of softwoods were now on a large scale, allowing pine, larch and the like to be used as minor structural timbers, reducing the demand for small-diameter oak, elm and ash. Thinnings from plantations could also supply such needs, much reducing the use of farmland timber for such purposes. More importantly, perhaps, it reflects the development of industrial sawmills, powered by water or steam, that allowed larger timber to be processed with relative ease, combined with the development of better roads, canals and ultimately railways, which allowed it to be transported more easily than before. The declining economic importance of bark after c.1860, as new forms of tanning using chromium salts replaced traditional methods, may also have been a factor, for small trees produce better-quality bark than larger ones and a greater quantity in proportion to their volume. Whatever the precise combination of causes, the change had a profound impact on the appearance of the countryside.

Conclusion

Major changes in England's trees and woodland thus occurred in the late eighteenth and nineteenth centuries, changes that were directly related to the profound economic and technological developments of the industrial and agricultural revolutions and the associated expansion in international trade. There was, in particular, a dramatic decline in pollarding and in the density of farmland trees in old-enclosed districts, for which the enclosure of the remaining open fields and commons and the associated spread of hedges and hedgerow trees provided only partial compensation. Overall, the number of trees in the countryside fell. There were also complex changes in the character of woodland, including the almost complete disappearance of wood-pastures, the widespread establishment of plantations without a coppiced understorey and the localised destruction of long-established coppiced woods. The area of woodland, broadly defined, in England as a whole probably rose by around 25 per cent in the late eighteenth and nineteenth centuries, and to some extent woods became more evenly distributed, with the greatest increases generally occurring in areas where they had formerly been thin on the ground – primarily areas of marginal land, especially moors and heaths. Lastly, there were marked changes in the management of timber, with significant increases in the age of felling and thus in the average age of tree populations both in woods and in the wider countryside. Yet, in spite of these upheavals, there were also marked continuities. Across most of England established coppices continued to be exploited and, in some districts, new ones established; oak, ash and elm continued to be the most common hedgerow timber; and in the majority of districts the countryside continued to be reasonably well treed, although still in relative terms poorly wooded. The pace of change was to accelerate markedly in the twentieth century, however, and it is to these more recent developments in our landscape that we must now turn.

 CHAPTER SEVEN

The threats to trees

Introduction

As we noted at the outset, underlying the historical enquiries described in this book are questions concerning the current condition of England's trees and woods. It is widely believed that we are losing trees from the countryside at a rapid rate as a consequence of agricultural intensification and industrial and urban expansion, and that many hundreds of ancient, semi-natural woods have been destroyed entirely or damaged beyond repair over the last century. Above all, it appears that our indigenous tree species are being assailed by a frightening succession of new diseases, mostly imported from abroad. All this is part and parcel of a wider and more general deterioration in the condition of England's natural environment. Many of these impressions are, we would emphasise here, at least partly true. But recent changes need to be carefully assessed and examined in historical context. We need to establish more clearly what precisely has changed over recent decades, by how much, and with what results. And we need to be clear about which particular social and economic developments have served to modify our tree populations, which were themselves largely shaped, in earlier periods, by social and economic processes and influences. Only by understanding such things can we begin to formulate a future for trees and woods in England.

One problem we have in thinking about the recent history of the environment is that it is often presented, in effect, as a 'game of two halves', divided by the upheavals of World War II. From the late 1870s farming began to slide into a long period of depression, principally caused by the expansion of the American railway network and the consequent conversion of the prairies of the mid-west into vast cereal fields. Britain, together with the rest of Europe, was flooded with cheap grain. No longer kept artificially high by the operation of the Corn Laws, repealed in 1846, wheat prices were halved between 1873 and 1893, while those for barley and oats fell by a third. Within a few years the problems of farmers were compounded by the arrival of cheap meat brought by refrigerated ships from the New World and Australia (Perren 1995; Perry 1974). The fortunes of farming recovered briefly during the First World War, but there was then a further slump, with only a partial and patchy recovery through the 1930s. According to many writers, because the countryside was now farmed at lower levels of intensity wildlife and a range of habitats benefited. Only in the period following the outbreak of the Second World War in 1939 did agriculture return to long-term profitability, as first the national government and latterly the European Economic Community and European Union introduced a range of subsidies aimed at increasing production. This ushered in a period of intensive farming that has continued, more or less, to the present day, and with calamitous results. Ancient pastures have been ploughed, wetlands drained, ponds filled in and woods, hedges and farmland trees removed on an awesome scale. This bipartite division is often used to frame discussions of environmental history. In John Sheail's words, '[w]hilst farming was generally depressed, the countryside of the first half of the century was typically diverse, beautiful and rich in wildlife. Farming boomed in the second half of the century, as those concerned with the conservation of amenity and wildlife … came close to despair' (Sheail 2002, 110).

The idea that the 'depression' years were broadly beneficial in environmental terms, in comparison with both what came after and before them, has also been expressed by Oliver Rackham when specifically discussing rural tree populations: 'The period 1870–1951 was, on the whole, an age of agricultural adversity, in which there was less money to spend on either maintaining or destroying hedges. Neglect gave innumerable saplings an opportunity to grow into trees' (Rackham 1986, 223). But the connections between nature and economics are seldom simple and this is perhaps especially the case with woods and trees.

The fate of farmland trees

There are, in fact, good grounds for believing that the numbers of farmland trees, which had been falling throughout the nineteenth century, did not increase, and in many areas continued to decline, during the first half of the twentieth. One key factor was the way that the agricultural depression impacted on the fortunes of large landed estates. Agricultural rents plummeted, especially in arable districts, and at the same time landowners were faced with a raft of other financial challenges, most notably Death Duties, introduced in 1894 and raised to 15 per cent by Lloyd George and subsequently, in 1919, to 40 per cent on estates valued at more than £200,000 (Thompson 1963, 325–30; Barnes 1984). Escalating financial difficulties led many to capitalise on their standing timber. In 1902 Rider Haggard noted the felling of hedgerow oaks in the area between Whissonsett and Wendling in Norfolk, commenting: 'I think that 'ere long this timber will be scarce in England' (Haggard 1902, 506). Lilias Rider Haggard similarly described in the 1930s how 'the wholesale cutting of timber all over the country is a sad sight, but often the owner's last desperate bid to enable him to cling to the family acres … ' (Haggard and Williamson 1943, 97). When, as was often the case, large estates were finally broken up in the first half of the twentieth century, the purchasers of particular farms – often their former tenants – were likewise keen to sell much of the timber, partly to help recoup the purchase price, partly to improve the yields in the adjacent fields. Where estates remained intact, moreover, it was often hard to find tenants in these difficult times, so that landlords were more sensitive than they had formerly been to the perennial complaints about the density of hedgerow timber. Lilias Rider Haggard described:

[a] consultation about the always difficult question of tree cutting on the farm. This particularly affects the arable fields, where the farming tenant has cause for some complaint. Decided somewhat sadly that some dozen small oaks must come out before the sap rises, or next autumn when the crops are off. (Haggard 1946, 73)

Forestry operations were increasingly concentrated in woods and plantations, where timber was also cheaper to extract – an important consideration, given that the rising scale of timber imports did nothing to improve the profitability of estate forestry.

Nor is there much evidence that saplings had a greater chance of growing into mature trees during the depression years in the manner that some have suggested. Hedges were still actively maintained in most districts, not least because outgrown hedges provided cover for the exploding population of rabbits. Indeed, descriptions of the countryside in the 1930s suggest that some hedges were already being grubbed out. Mosby noted how, in north-east Norfolk, there was 'a tendency in some areas to enlarge the fields by removing the intervening hedge. Where this has been done the farmers, particularly those who use a tractor plough, have reduced their labour costs' (Mosby 1938, 203–4). Butcher in 1941 described much of the Suffolk landscape as having 'Hedges … kept as low as possible or even rooted out. Consequently one characteristic of the district is the hedgeless or almost hedgeless fields surrounded by deep ditches' (Butcher 1941, 357). Not surprisingly, the naturalist George Bird was able to bemoan – in a lecture on Suffolk birds delivered in 1935 – the 'devastation of the countryside by loss of timber and hedge-shelter' that had occurred during the previous decades (Bird 1935). Elsewhere in the arable east the situation was similar. Scarfe in 1942 described north-west Essex as a land of 'low hedges, large fields and

7.1 A photograph taken in the 1920s showing large arable fields in west Suffolk. A single oak tree, a former pollard, has been left after hedges have been grubbed out: the simplification of the East Anglian landscape was already under way.

7.2 One of the 'pine rows' which are a typical feature of the landscape of the East Anglian Breckland. Originally a hedge, planted in the early nineteenth century, it was allowed to develop into a line of trees during the agricultural depression of the early twentieth century.

few hedgerow trees' (Scarfe 1942, 463); in the Kelvedon district 'the hedges are low and well trimmed and there are few hedgerow trees'; along the western fringes of the county much of the land had 'no hedges or trees and … large fields' (Scarfe 1942, 437); while the district around the Rodings was 'a land of few hedges and fewer trees' (Scarfe 1942, 440) (Figure 7.1). And yet the extent to which farmland trees and, in particular, hedgerows were lost in this period, even in eastern arable districts, should not be exaggerated. Changes to the fabric of the countryside were minor compared with what was to come in the post-war years. Moreover, in western pastoral districts the loss of trees and hedges was on a more limited scale, although even here most field boundaries continued to be well maintained and there is little evidence to suggest that tree populations were significantly augmented with self-seeded specimens. Indeed, only in a few limited areas does the neglect of hedges really appear to have led to a significant rise in the numbers of farmland trees. The most striking example is the East Anglian Breckland. Here, enclosures made from heaths and open fields in the early decades of the nineteenth century were often surrounded by hedges of Scots pine, most of which are still shown as hedges, rather than as lines of conifers, on the 1880s Ordnance Survey 6-inch maps. Management declined during and after the First World War, and by 1946 – when the RAF produced their comprehensive aerial survey of Britain – most had become the lines of picturesquely twisted trees that remain an icon of Breckland (Barnes and Williamson 2011, 138–52) (Figure 7.2).

It is difficult to quantify the extent to which farmland trees were lost during the first half of the twentieth century. Comparison of the two obvious sources – the Ordnance Survey 6-inch maps of the 1880s and 1890s and the RAF vertical air photographs of 1945–7 – are fraught with problems. It is hard to count individual trees on the latter source and harder still to distinguish mature specimens from the kinds of 'sapling' ignored by the Ordnance Survey.[3] This said, comparison does seem to suggest a continuing downward trend. On the boulder clays of East Anglia, for example, average densities of around five trees per hectare in the 1880s may have fallen, by 1946, to around four; on the dipslope of the Chiltern Hills average densities of around four trees per hectare in the 1880s had declined, by 1946, to around three; while on the heavy clays in south Hertfordshire and south Essex the apparent decline was from 4.5 to less than four.

While the years of depression may have witnessed a continuing reduction in the numbers of farmland trees in most if not all areas, the revival of farming fortunes through the 1960s and 1970s certainly saw more rapid attrition. Increases in farm size, the widespread adoption of tractors and combine harvesters and a shift in eastern districts away from mixed farming to purely arable husbandry all contributed to a wholesale loss of hedges and hedgerow trees, as well as to the grubbing out of many long-established coppiced woods. Between 1946 and 2000 around 150,000 miles – 240,000 kilometres – of hedges were destroyed in England and Wales, with the greatest losses occurring in the eastern counties (Jones 2012; Baird and Tarrant 1970). Trees were sometimes allowed to remain when hedges were grubbed out, but they were usually felled. All this devastation was compounded by the arrival in Britain, in the late 1960s, of Dutch elm disease. From a small number of initial points of infection this spread at a rate of around eight miles (13 kilometres) a year, and by the late 1980s the disease was well entrenched throughout the country, leading to the almost complete destruction of elm as a tree although not, of course, as a suckering shrub (Rackham 1986, 240–47; Brasier and Gibbs 1973).

But, just as the benign effects of the depression years can be exaggerated, so too can the extent of the decline and destruction which followed them in the post-war period. Hedgerow destruction had, in fact, begun to tail off by the 1970s, and was made more difficult by legislative changes in 1997. Through the later 1970s and 1980s agricultural fashions began to move away from a single-minded desire to maximise food production and towards the combination of farming with the achievement of at least some environmental and amenity benefits. Following the publication of the Countryside Commission's *New Agricultural Landscapes* report in 1974 (Westmacott and Worthington 1974) novel forms of rural planting started to appear, intermediate between woods and hedgerows, especially small groups of trees in field corners. From the 1980s government agri-environment schemes – especially Countryside Stewardship – encouraged the replanting of hedges and hedgerow trees, as well as the establishment of further clumps and narrow belts in the countryside. Even without financial inducements, small areas of woodland were often

---

3   The second and provisional editions of the Ordnance Survey 6-inch survey did not record farmland trees.

planted in order to provide cover for game and to compensate for the loss of elm. And to this we need to add the narrow strips and lines of trees planted by local authorities and other public bodies to screen new roads and developments. The extreme destruction of the post-war period, in other words, did not last long, and was succeeded by a period in which, in some places at least, landowners began to replant, if usually on a modest scale.

Some studies suggest that the numbers of farmland trees have continued to decline over the last few decades, but the figures supplied by the Forestry Commission appear to show that trees growing outside woods have actually *increased* significantly in many if not all English counties since the 1970s. In Norfolk, for example, it is claimed that there are now on average 4.7 such trees per hectare, a figure which is actually *above* the average density apparently shown on the late nineteenth-century Ordnance Survey maps (Forestry Commission 2002). In Northamptonshire, to take another example, the figure is 2.1 per hectare, in this case probably below that for the late nineteenth century, although not by a very significant margin. Such increases have not occurred in all counties, but measured nationally there appears to have been a significant upward trend. On the face of it, the corner has been turned, largely as a consequence of large-scale planting by farmers and others. In reality, the increases apparently indicated by the official figures are not only or perhaps even primarily the consequence of deliberate planting, but of changes in land use and land management. And whether the new trees have really compensated for those lost earlier in the twentieth century is debatable.

Much depends – as so often in the study of trees – on what we measure and how; and, in particular, on how we regard trees growing in small clusters on farmland, in what the Forestry Commission defines as 'groups' (covering less than 0.1 of a hectare) and as 'linear plantings' (i.e., narrow strips or belts). The latter are of particular importance because it is the marked rise in their numbers that accounts for most of the officially recorded increases in non-woodland trees. Most such features are less than 15 metres in width and a significant proportion comprise single lines of trees. While some represent deliberate planting, especially alongside new roads, most seem to have developed organically. Many, for example, occupy the lines of former railways closed as a consequence of Beeching's disastrous programme in the 1960s. More importantly, a very high proportion mark the lines of outgrown hedges. Well into the twentieth century there was a fairly clear distinction between a hedgerow tree and the hedge in which it grew, because hedges were generally kept low by plashing or cutting. In the post-war period, and especially from the 1970s, this distinction began to blur in many areas, as hedges were allowed to develop into lines of close-set trees, sometimes simply through neglect but sometimes as a consequence of well-intentioned attempts to increase tree numbers (Figure 7.3). Hedges containing a significant proportion of maple and ash were particularly likely to develop in this manner. Even where hedges have not fully developed into such lines, a decline in regular cutting has often allowed a proportion of their faster-growing constituent plants (again, especially ash and maple) to become trees.

7.3 The increase in farmland tree numbers in the second half of the twentieth century recorded in many official statistics is in part the consequence of the development of unmanaged hedges into lines of close-set trees, as here in mid-Norfolk.

In this context, it is noteworthy that the counties in which farmland trees have witnessed the most significant recovery since the 1970s are, for the most part, those dominated by arable agriculture, where livestock are absent and hedges are therefore no longer expected to serve any useful function. Such circumstances have also favoured the regeneration of small 'groups' of trees around pits and ponds, and beside watercourses, in fields no longer grazed by sheep or cattle, together with the regeneration – or deliberate planting – of angles and corners difficult to plough. In spite of a steady reduction in the number of free-standing hedgerow trees, a substantial rise in average field size and hedgerow removal on a massive scale, such developments have often significantly increased the numbers of non-woodland trees in arable areas. The number of individual specimens crammed along the line of an outgrown hedge is far higher than that found when individual trees are thinly scattered along one that is still managed.

In Norfolk, an intensively arable county, there was a 21 per cent reduction in the numbers of traditional free-standing farmland trees between 1970 and 1998, but this was more than compensated for by a 120 per cent rise in the number growing in 'groups' and a 293 per cent increase in those growing in 'linear features', resulting in an overall expansion in the numbers of non-woodland trees of 120 per cent. In Suffolk, similarly, the number of non-woodland trees rose by 100 per cent across this same period, in spite of a 24 per cent decline in the number of free-standing trees, because those growing in 'groups' and 'linear features' increased by 44 per cent and 240 per cent respectively. Most dramatically of all, in Lincolnshire – England's most intensively arable county – the number of free-standing trees declined by 33 per cent, but the figure for non-woodland trees as a whole rose by 194 per cent, largely because numbers growing in 'linear features' increased by a staggering 1,122 per cent. Of course, small groups and linear features also increased through deliberate planting, and these counties were among those particularly targeted for government agri-environment schemes. In addition, the sheer scale of the reduction in tree numbers in the more intensively arable regions ensured that deliberate planting as part of agri-environment schemes or for sporting purposes has had a greater impact, in percentage terms, than in regions where post-war destruction was less. But changes in agricultural practices and land-use patterns appear to have been the main influence. In this context, it is noteworthy that in many northern and western areas numbers of non-woodland trees have continued to decline. Hedges often (although not always) continue to be maintained in livestock-farming areas, while in the uplands the fields are frequently bounded by drystone walls that are unable to develop, through neglect, into lines of trees. This said, in many livestock areas hedges have also tended to develop into lines of trees, barbed wire fences running along their length now serving as the principal barrier to movement.

Some of the complexity of recent changes in farmland tree populations can be illustrated by looking in detail at particular local areas. Henry Keymer surveyed a farm of 80 acres (32 hectares) in the Norfolk parish of Scarning in 1764 and recorded a total of 1,095 trees, or 31 per hectare (NRO BCH 20). The first edition Ordnance Survey 25-inch map of 1886 marks only 72 free-standing trees, a total which had fallen to around 55 by 1946, to judge from the RAF aerial photographs. Today there are only 32 free-standing hedgerow trees within the area: one per hectare. However, when the abandoned railway line that runs through the middle of the area was turned into a new route for the A47 in the 1970s narrow strips of trees were planted to either side, the widths of which – c.13 metres – would allow them to be classified as 'linear features' in Forestry Commission terms. These contain, within the area mapped by Keymer, a further 514 trees. In addition, small 'groups' of trees have developed around field ponds, further increasing the overall total. Looked at in one way – considering only free-standing hedgerow trees – the area thus saw a catastrophic decline of 93 per cent in tree numbers in the course of the late eighteenth and nineteenth centuries, followed by a more muted reduction of 55 per cent during the twentieth. But, looked at in another, the later twentieth century has witnessed a significant recovery within the area, so that the number of non-woodland trees is now greater than it was in the late nineteenth century, at 554 as against 72, although this is still only around half the 1764 level. However, another Norfolk property surveyed by Keymer, in Beeston-next-Mileham, although lying only seven kilometres to the north-west, exhibits a very different pattern (NRO WIS 138). On the enclosed portion of the farm, covering 25.5 hectares, there were 531 trees in 1764, or 21 per hectare. By the 1880s this had been reduced to around 100, in part through wholesale boundary reorganisation, and by 1946 to a mere 60. Today there are only seven free-standing trees within the plots mapped by

Keymer. Here, moreover, 'small groups' around ponds and the like raise the overall total to only around 20, and there are no 'linear features'.

Such marked variations over short distances are evident elsewhere. In east Hertfordshire an estate at Whempstead in Little Munden apparently had 653 mature trees on 98 acres when mapped in 1808, including examples growing in well-timbered fields or diminutive wood-pastures – an average of roughly 16 per hectare (HALS 81750). By the 1890s this had been reduced, within the bounds of the same area, to around 60 trees. Today, in spite of much field amalgamation, there are still 55 free-standing trees, several of them only just mature; but if we include examples growing in 'groups' and 'linear features', especially those in a magnificent outgrown hedgerow, brimful of mature ash and maple, the figure rises to 130, significantly above the number present in the late nineteenth century. Some four kilometres to the south-east the central portion of a farm at Green End, Standon, covering some 40 hectares, displays a rather different pattern. Here, to judge from a detailed estate map, there were around 510 trees in 1774 (HALS E/2833). By the 1890s this had fallen to 120. But today, within the same area, there are actually more free-standing trees – around 130 – nearly a third of which are less than *c.*35 years old. The contents of 'groups' and 'linear features' brings the total number of trees to over 150. Yet, just two kilometres to the east, on 32 hectares forming what was part of the demesne land of the manor of Barwick, the situation is different again. In this case, 75 hedgerow trees are shown (with apparent accuracy) on a map of 1774 (HALS E/2832), a number that had declined to 56 by the 1880s. Today there are only 32 mature trees here. In this case no 'linear plantings' or 'groups' exist, and most of the trees are mature or over-mature. The contrasts exemplified by these three cases are between a farm run with a keen interest in conservation (Green End, Standon); one on which arable production appears the main or only concern (Barwick); and one lying in an intermediate position on this notional spectrum, in which remaining hedges have been allowed to develop into broad bands of shrubs and trees (Whempstead).

One effect of recent developments, and especially of the practice of allowing hedgerow shrubs to grow into trees, has been to change the species composition of farmland populations, with an increase, in particular, of maple (which now accounts for 9 per cent of the trees at Green Farm and 36 per cent of those at Whempstead) and ash. Where hedges are hacked back or grubbed out and little new planting has taken place – as at Barwick and Beeston – oak tends to be more dominant, usually comprising well over half of farmland trees, although these are mostly over-mature specimens.

In short, the suggestion that the countryside has continued to lose trees at an unremitting rate since the Second World War is only partially true. There is much variation from district to district, even from farm to farm, but to a surprising extent (and depending on what, precisely, we include in our calculations) the absolute numbers of farmland trees in most southern and Midland districts has probably increased significantly since the late 1970s. Whether these are the right *kinds* of tree, planted in the right configurations and locations, is a more difficult matter, to which we shall return.

The expansion of woodland

Just as most people perhaps assume that trees are disappearing at a steady rate from the countryside, many probably believe that the area under woodland in England has dwindled markedly over the last half century or so, giving way to farmland, housing and industry. In one important but limited sense this is true. The area occupied by ancient coppiced woodland has indeed contracted, and little of what remains is still managed intensively, to encourage the particular range of plants and invertebrates associated with this important habitat. But the area of woodland *of all kinds* – including plantations – has not fallen. It increased markedly in the course of the twentieth century, and it continues to rise.

To an extent this is the consequence of the activities of the Forestry Commission. This body was established in the immediate aftermath of the First World War, during which enemy blockade had starkly revealed both the relative deficiency of timber in Britain and the extent of its reliance on imports. There had, in particular, been an acute shortage of pit props, which seriously threatened the war effort. In 1916 the prime minister, H.H. Asquith, appointed the Forestry Sub-Committee of the Ministry of Reconstruction 'To consider and report upon the best means of conserving and developing the woodland and forestry resources of the United Kingdom, having regard to the experience gained during the War' (Ryle 1969, 25). The government also had other objectives in stimulating

7.4 Workers planting a Forestry Commission plantation, by hand, in the 1930s.

forestry, most notably the alleviation of rural unemployment (Sheail 2002, 84–90). The committee proposed that, over the following 80 years, no fewer than 1,770,000 acres (*c.*72,000 hectares) of land should be planted with trees. In 1919 the Forestry Act established the Commission, which immediately began to acquire land, most of it from hard-pressed landowners desperate to sell or lease at this time of agricultural adversity (Ryle 1969, 25–39). The main areas targeted were marginal land, especially heaths (or recently abandoned farmland earlier reclaimed from heath) and upland moors – thus continuing and intensifying the pattern established during the eighteenth and nineteenth centuries. Large areas of heath and derelict arable were acquired and planted during the 1920s and early 1930s, especially in the East Anglian Breckland, on the coast of Suffolk and in parts of Hampshire, Dorset and Nottinghamshire; and extensive tracts of upland moor, particularly in Yorkshire and Northumberland, were acquired mainly during the 1930s. But other areas were also planted by the Commission, most notably the Crown woodlands, much of which had been retained following the enclosures of the royal forests in the nineteenth century. The speed of planting, all undertaken by hand in the inter-war years, was remarkable (Figure 7.4). In 1927 around eight million trees were established on 3,700 acres (1,500 hectares) in Breckland alone (Dannatt 1996; Forestry Commission Annual Reports; Forestry Commission *Working Plan for Thetford Forest*, 1959) (Forestry Commission archives, Santon Downham). By 1950, in England as a whole, no fewer than 302,000 acres (*c.*122,000 hectares) of new forest had been planted by the Commission (Ryle 1969, 298–9).

The new plantings were overwhelmingly coniferous, and many existing areas of deciduous woodland, including many ancient coppiced woods, were also replanted with conifers by the Commission, although mainly in the post-war period. In southern districts, on sandy soils, Scots pine was the tree of choice in the 1920s, although it was increasingly replaced through the 1930s and 1940s by Corsican pine, which is faster-growing and more disease resistant. Small areas of Douglas fir, European larch, silver fir, western red cedar, western hemlock, maritime and lodgepole pine were also established. In northern and western areas sitka spruce, Norway spruce, European larch, Japanese larch and lodgepole pine were mainly employed. Some of the initial plantings, in lowland areas especially, also included a significant proportion of indigenous hardwoods such as oak and beech. In 1935, for example, 1,186 acres (480 hectares) of conifers were planted in the East Anglian Breckland and as many as 428 acres of hardwood trees, principally oak and beech (Skipper and Williamson 1997, 30–31). But on such poor soils indigenous trees grew more slowly than conifers and were more vulnerable to drought and deer, and their planting soon declined. In Breckland in particular, however, some continued to be established as belts along the sides of roads, in part as a fire prevention measure and in part to assuage growing opposition to the 'serried rows of conifers' that were transforming the formerly open landscapes of heaths and 'brecks' (Skipper and Williamson 1997, 65–7; Chard 1959). In 1939 Julian Tennyson described how:

7.5 Aerial view of Mortimer Wood, a Forestry Commission plantation established on former moorland on the Herefordshire/Shropshire border.

> [t]he Commission has worked its way steadily through the centre of Breckland, buying and leasing estates, removing boundary after boundary, until now there is scarcely a couple of miles of ground left unplanted between Lakenheath Warren and Elveden in Suffolk and the road from Methwold to West Wretham in Norfolk. It has swept everything before it: the heaths and brecks in its paths have disappeared for ever. Small wonder that those who loved the old spirit of Breckland should complain that they can now scarcely even recognize their own country (Tennyson 1939, 76–7).

Indeed, at a time when more and more people believed that the countryside was under significant threats from modernity, the vast new plantations were almost everywhere greeted with horror (Figure 7.5). The Commission's plans to plant 300 hectares in Upper Eskdale in the Lake District in the 1930s led to a prolonged campaign that culminated in a successful public subscription to raise the £2 per acre compensation required for not planting the land, much of which subsequently passed to the National Trust. The dispute also led to a voluntary agreement negotiated between the Commission and the Council for the Protection of Rural England which restricted the establishment of plantations within a central block of *c*.300 square miles of the Lakes (Crosby and Winchester 2006, 236–7; Simmons 2003, 156–7).

The sudden imposition of extensive tracts of conifers on heaths and moors certainly had, in the medium term at least, negative effects on the local flora and fauna. The dense conifers shaded out the plants beneath, and their needles blanketed the ground, further increasing the acidity of soils already base-poor. Plants such as heather and gorse often survived in the rides, but many characteristic species of moor and heath were lost. The forests were, indeed, an unnatural imposition on the landscape and, not surprisingly, vast areas planted with a limited range of species were vulnerable to pathogens. In Breckland large areas soon suffered attacks from the pine shoot moth *Evetria buoliana* (Ross 1935; Gibbs *et al.* 1996). More serious was the emergence, as thinning began in the late 1930s, of the fungus *Fomes annosus* (since renamed *Heterobasidion annosum*) (Day 1948; Rishbeth 1963; Wass 1956). The airborne spores took hold in the stumps of felled trees and then spread through the root system to unfelled specimens. Infected stumps were initially isolated by being surrounded with a trench and then inoculated using the fungus *Peniophera gigantean*, but the disease continued to affect new planting, especially on the more calcareous soils, and since the 1970s stumps in such areas have been routinely bulldozed into long rows to prevent the spread of infection (Skipper and Williamson 1997, 72–4).

The Forestry Commission looms large in histories of the twentieth-century landscape, perhaps in large measure because its activities were targeted at many of the more valued landscapes in England, in terms of recreation, visual beauty and natural history. Most districts were unaffected by the Commission's activities – and yet they, likewise, usually saw a significant increase in woodland area through the inter-war years. Moreover, the large-scale planting of new land by the Forestry Commission largely came to an end with the Second World War, but in the second half of the twentieth century the area occupied by woodland continued to expand, often at a greater rate than before. While the activities of the Commission were clearly a major influence on the development of woodland in England in the twentieth century, other factors and processes were evidently at play.

One was deliberate planting by bodies or individuals other than the Commission. Although, as we have seen, many landed estates were broken up in the first half of the century, large numbers survived and their owners continued, as they had long done, to indulge in planting schemes. During the years of the agricultural depression plantations represented a form of agricultural diversification on land that could no longer be profitably farmed, and also provided cover for game. In the post-war period private planting continued, but was now often carried out by a new breed of large landowner. In the 1950s and 1960s the increasingly capital-intensive character of agriculture saw the emergence, in arable districts especially, of larger and larger farms, as successful individuals bought up the holdings of their neighbours. Private owners of all kinds were now also motivated by a desire to improve the appearance and the countryside, and to benefit its wildlife, and from the 1970s they were often supported in such endeavours by government-funded agri-environment schemes. In addition to all this, in the post-war period much planting was undertaken by public bodies, such as county and district councils, to screen new developments or to improve the appearance of abandoned mine workings and the like.

From 1972 the Woodland Trust began to acquire and plant significant areas with trees, and continues to do so, often on an extensive scale. Most of this post-war conservation and amenity planting was of deciduous hardwoods, it should be noted: and, from the early 1980s, the emphasis of the Forestry Commission itself changed towards deciduous planting and in favour of multi-use forestry with an important recreational and conservation element. As a result of these changes the proportion of purely coniferous woodland began to decline through the second half of the twentieth century at the same time as the overall area of woodland continued to rise.

A second way in which woodland increased during the twentieth century was through natural regeneration on various kinds of ungrazed and derelict land, something that, likewise, largely involved indigenous trees rather than conifers, and especially silver birch, previously a relatively rare species in most lowland areas. In many districts significant tracts of common land had escaped the enclosures of the eighteenth and nineteenth centuries and already, by the 1860s and 1870s, some were being regarded by local residents in urbanising and suburbanising areas as spaces for recreation: hence the establishment of the Commons Preservation Society in 1865, which was largely responsible for preserving Hampstead Heath, Epping Forest and other London commons from being built over (Cowell 2002, 148–9). In addition, even where commons had been enclosed by parliamentary acts the local poor had often received a 'fuel allotment' where they could continue to cut peat or gorse, and in some cases graze some livestock. All these areas of rough land came to be used less and less intensively from the later nineteenth century. As coal came to be the principal domestic fuel, peat and gorse remained uncut: as early as 1875 it was reported that the local population had ceased to cut turfs on Whitwell Common in Norfolk because 'the houses and fireplaces of the commoners are unsuitable for the burning of turf' (Birtles 2003, 209). Grazing on allotments and commons also declined. Few members of the rural poor could afford a cow, while the larger farmers tended to shun the rough grazing provided by common land. Through the 1920s and 1930s the escalating scale of car traffic (most commons in lowland England were crossed by public roads) was also causing difficulties, not least because motorists often left open the gates to the common, allowing sheep and cattle to wander off down the lanes. Naturalists were already commenting on the increasingly scrub-covered character of commons and similar land in the first half of the twentieth century. In Norfolk, for example, many were – by the time of the First World War – already being invaded by scrub, 'dominated by whitethorn, blackthorn, furze, and *rubus*' (Clarke 1918, 305). Birch and pine, and sometimes oak, were also colonising: their seedlings were 'common on many heaths' (Clarke 1918, 308). But it was in the middle and later decades of the century, to judge from the evidence of Ordnance Survey maps, that many of these areas became truly wooded.

Woodland developed spontaneously on other kinds of under-used or derelict land. The twentieth century saw a steady increase in the area occupied by gravel pits, quarries and other mineral workings, and when these were abandoned they often came to be occupied by trees. Old spoil heaps and slag heaps were often colonised by secondary woodland, as were a wide range of other types of industrial site rendered redundant by the complex economic changes of the twentieth century, especially in northern districts. All these various areas had in common the fact that they were not systematically grazed by livestock, as almost all available land would have been in the seventeenth and eighteenth centuries. To some extent, the presence of rabbits reduced the speed with which woodland regenerated, but the great myxomatosis outbreak of the 1950s led to a massive reduction in their numbers, substantially increasing the spread of bushes a nd trees.

The speed and scale with which woodland expanded is difficult to chart with precision owing to the fact that different methodologies and definitions have, over time, been employed by successive official surveys. If we define a 'wood' as an area of trees occupying more than around 0.5 hectares then we might reasonably estimate that – in spite of a century or more of extensive planting by landed estates – woods and plantations still occupied a mere 5 per cent of England's surface area in 1895. This proportion does not appear to have risen significantly over the following decades. Estimates made in 1905 put the figure at 5.2 per cent and in 1913 at 5.2 per cent, and woodland was reduced during the First World War, when as much as 180,000 hectares were clear-felled and not immediately replanted – although how far this really constitutes woodland 'loss' is a moot point, given that in most cases these areas eventually returned to a wooded state (Board of Agriculture Returns, 1896, 36; Board of Agriculture Statistics, 1910, 56; Board of Agriculture Stats, 1914, 92). Woodland expanded steadily through the 1920s and 1930s, however, and by the end of Second World War occupied around 5.8 per cent of England's land area. Growth then accelerated, reaching 6.8 per cent of land area in 1965, 7.3 per cent in 1980 and 8.4 per cent by 1996. Today it stands at a little under 10 per cent (Forestry Commission 2016, 5). In broad terms, and contrary to what many may assume, the area under woodland has thus roughly doubled since the end of the nineteenth century. As we noted, it is possible that the scale of this increase has been exaggerated by differences in the minimum defining size of a 'wood' employed in different surveys. More importantly, as always, overall national figures hide a mass of regional variation, and especially so when twentieth-century developments are viewed in longer chronological perspective. Different counties and different localities have their own distinct patterns of recent woodland history that remain to be fully explored.

In well-wooded Hertfordshire, for example, around 7.5 per cent of the land area was occupied by woodland and wood-pasture in the later eighteenth century, perhaps falling to around 6.3 per cent by 1895 (Board of Agriculture Returns, 1896), although, as noted (above, p. 73), the first of these figures may not be entirely reliable. Here there seems to have been only slow growth in the first half of the twentieth century, in part a reflection of the fact that there was little Forestry Commission activity in the county. In 1905 woodland cover was estimated as 6.5 per cent of the land area, in 1913 at 5.2 per cent, in 1924 at 6.4

per cent and in 1949 at 6.6 per cent. By 1959, however, a decisive increase had occurred, to 7.5 per cent; by 1980 the figure was 7.7 per cent; and by 1998 it had reached around 9 per cent (Board of Agriculture Statistics, 1910, 56; Board of Agriculture Statistics, 1914, 92; Forestry Commission 4th Annual Report, 1924, 25). Today it is probably over 10 per cent (www.natureonthemap.naturalengland.org.uk). This is in spite of the fact that in the course of the twentieth century the south of the county experienced large-scale suburbanisation. Much of the increase was a consequence of amenity and conservation planting associated in part with the establishment of planned 'Garden Cities' and 'New Towns', but some of it had other causes. A surprising number of large commons in Hertfordshire escaped parliamentary enclosure and, by the 1960s, many were regenerating to secondary woodland. In the Chiltern Hills Berkhamsted Common, Pitstone Common and Aldbury Common are still shown as mainly open ground on Ordnance Survey maps from the 1960s, but are now mainly wooded; the same is true of most of the smaller commons scattered across the Chiltern dipslope, such as Chipperfield Common, Bernards Heath in St Albans and, to a lesser extent, Nomansland Common, on the boundary between St Albans and Wheathampstead. These examples alone would have served to increase the amount of woodland in the county by some 600 hectares, nearly 0.4 per cent of its total land area. The regeneration of woodland on abandoned common land and other rough grazing produced even more significant increases in other counties in southern England, where greater expanses of common land had escaped enclosure. In Hampshire, for example, the bushing-over of extensive commons largely accounts for the remarkable rise in woodland from around 12 per cent to over 17 per cent of county area between 1895 and 2000.

The Midland county of Northamptonshire displays a very different pattern. Here, as we have seen, the proportion of the land area occupied by woodland fell drastically with the enclosure of the royal forests, from perhaps 7 per cent of the land area in the mid-eighteenth century to as little as 4.5 per cent in 1895. As in Hertfordshire, woodland area then remained little changed throughout the first half of the twentieth century, estimates made in 1895, 1905, 1913, 1924, 1939 and 1947 all suggesting that between 3.9 and 4.5 per cent of the county was occupied by woods and plantations (Board of Agriculture Returns, 1896, 36; Board of Agriculture Statistics, 1910, 56; Board of Agriculture Statistics, 1914, 92; Forestry Commission 4th Annual Report, 1924, 25; Forestry Commission 15th Annual Report, 1935; Beaver 1943, 362–5). Some new areas of woodland were certainly planted by large estates in this period – the records of the Boughton estate, for example, show an active interest in forestry in the early years of the twentieth century, with 1,000 oak, 100 ash, 200 larch, 1,150 beech and 200 sycamore being planted in 1908 alone (Boughton House archives). But woods were also lost, not least as a consequence of the development of ironstone quarrying from the late nineteenth century, much of which was concentrated in the area of Rockingham Forest, around Kettering and Corby (Orchard 2007). New planting and the destruction of existing woods more or less cancelled each other out for much of the century and, in 1979, according to another Forestry Commission survey, the total area of woodland within the county was still virtually unchanged, at 11,600 hectares, or 4.5 per cent of the land area – a pattern of stasis that probably reflects in part the paucity of surviving common land in the county across which woodland could regenerate. By 2000, however, woodland area had increased significantly, to 14,500 hectares, around 5.7 per cent of the land area, and it is now probably around 7 per cent (Forestry Commission 2002; www.natureonthemap. naturalengland.org.uk). This expansion was again a result of grant schemes, countryside stewardship agreements and game and amenity planting, together with much replanting and regeneration over abandoned industrial sites, especially derelict ironstone quarries. The area under woodland in the county has thus probably almost returned to that of coppiced woods and wood-pastures combined in *c.*1750 – a remarkable recovery.

Other counties displayed radically different patterns of long-term change. In Norfolk, for example, where extensive tracts of poor sandy land exist, the area of woodland increased significantly during the nineteenth century, as we have seen, from around 2.5 per cent of land area in 1797 to 4 per cent in 1895 (Board of Agriculture Returns, 1896, 36). In 1905 it was estimated that woodland accounted for 23,926 hectares, or 4.5 per cent of the land area (Board of Agriculture Statistics, 1910, 56); in 1913 it was thought to be very slightly less, at 23,523 hectares (Board of Agriculture Statistics, 1914, 92); and in 1924 the Forestry Commission reckoned it as 19,798 hectares, or 3.7 per cent (Forestry Commission 4th Annual Report, 1924, 25). These differences probably reflect variations in survey methods, and it was only during subsequent years that the area under woodland appears to have increased significantly, the Forestry Commission estimating a total of 27,256 hectares (5.2 per cent of the county's land area) in 1932–5 and the Land Utilisation Survey in 1938 no fewer than 33,144 hectares, or 6.3 per cent (Mosby 1938; Forestry Commission 15th Annual Report, 1935). This increase was largely a consequence of the vast plantations established by the Forestry Commission in Breckland, but also resulted from the regeneration of woodland on commons, poors' allotments and other rough ground. Clarke in 1918 had estimated that such land covered some 11,324 acres (4,582 hectares) in the county (Clarke 1918, 295), but this excludes most low-lying fen meadows and reed beds, especially in the valleys of the Broadland rivers, where there was also much development of secondary woodland in the middle decades of the twentieth century. Already, by 1939, Boardman was able to describe how:

> A great change has taken place over a large portion of the river valleys about here. Thirty years ago all the large acreage of rough marsh covered with sedge rush and grass was mown for what was called marsh litter. The best was made into hay, or chaffed and used for feed locally or sent to London for bus horses and cows, while the rougher stuff was put into the bullock yards to be trodden into manure. Now, since acres upon acres of this material remain uncut and the vegetation gets into such a terrible tangle, the marshes have to be burned. Alders, birch and sallows are taking possession … . When the marshes were mown regularly all the young trees were kept under. (Boardman 1939, 14)

Woodland continued to expand in the post-war years, and by 1980 42,697 hectares of woodland were recorded in the county (8 per cent of the land area). By 1997 the figure had reached 52,046 (9.7 per cent) and today it stands at over 10 per cent. Once again, this increase reflects a combination of factors: the continued establishment of numerous small woods by landowners for conservation and sporting purposes, the planting activities of local authorities and public bodies and the continuing process of natural regeneration on ungrazed derelict and marginal land.

Yorkshire provides a good example of developments in northern and upland districts. Here, as in Norfolk, major increases in the area under trees occurred in the first half of the twentieth century as a result of the activities of the Forestry Commission, mainly on the margins of the North York Moors. Planting began in 1924 and by 1938 no fewer than 9,010 acres (3,246 hectares) of new woodland had been established, with the largest concentration in Allerston Forest, now known as Dalby Forest (2,209 hectares), in the Hambleton Hills (Wooldridge 1945, 383–4). After the Second World War, in contrast to the situation in Norfolk, there was a further expansion in the Commission's holdings, to more than 19,000 hectares, most of which were likewise concentrated around the margins of the North York Moors (Forestry Commission 2002). Although, as in other areas, some contribution was made by the spontaneous regeneration of woodland (especially on abandoned industrial sites) and the planting of private landowners, the Forestry Commission plantations probably account for most of the increase in woodland area in the first two-thirds of the twentieth century. Before the commencement of planting in the mid-1920s the overall woodland total for the county appears to have remained relatively stable, variously estimated at around 3.5 per cent of land area in 1895, 3.8 per cent in 1905, 3.7 per cent in 1913 and 3.6 per cent in 1924; but it then rose significantly, to 4.2 per cent, by the mid-1930s (Board of Agriculture Returns, 1896, 36; Board of Agriculture Statistics, 1910, 56; Board of Agriculture Statistics, 1914, 92; Forestry Commission 4th Annual Report, 1924, 25; Forestry Commission 15th Annual Report, 1935). Subsequent comparisons of woodland areas are difficult to make owing to changes in administrative boundaries, but by 1980 the figure had probably risen to above 6 per cent of the old area of the three Ridings; by 1998 it was around 7 per cent; and it is now 7.2 per cent, a total of 113,328 hectares (Forestry Commission 2002; www.natureonthemap.naturalengland.org.uk).

Many factors thus influenced the extent to which the area of woodland increased in different parts of the country in the course of the twentieth century, but perhaps the most important was land quality: in areas of good agricultural (and especially arable) land, woodland area did not increase significantly and in some areas continued to contract. It was on poorer land, as in the nineteenth century, that most new woods were planted; and it was on such land, especially in the south, that many areas of woodland developed through natural regeneration, as commons and poors' allotments ceased to be exploited for grazing or fuel. Indeed, whereas in the nineteenth century the expansion of woodland was generally greatest in the most poorly wooded districts, in the twentieth century this connection

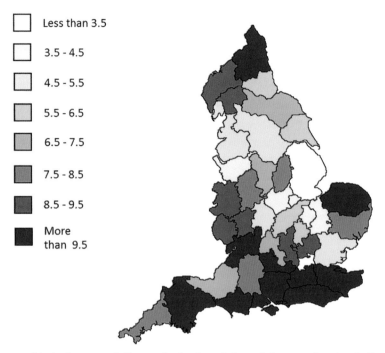

7.6 Percentage of the land area occupied by woodland at the end of twentieth century (based on the Forestry Commission Inventory for 1998 and adjusted to the boundaries of the old counties, for comparison with Figure 6.9).

was broken owing to the scale of natural regeneration. Surrey was thus already in 1895 the third most densely wooded district in England, with 11.8 per cent of its area occupied by woods and plantations, but by 2000 this had nearly doubled, to 22.4 per cent, largely although not entirely as a consequence of the bushing-over of the great heaths that had also characterised the county. This said, woodland distribution was structured by a host of other factors, including landownership and the extent of industrial expansion and decline, and patterns and relationships are difficult to chart in part because statistics have usually been recorded on the basis of counties, artificial administrative units that resolutely cut across the boundaries of farming districts and landscape regions: a county such as Buckinghamshire, for example, has a well-wooded southern portion, embracing part of the Chiltern Hills, and a more sparsely wooded northern part, extending far into the former 'champion' Midlands. The situation is not helped by the fact that the boundaries of local government units have changed over time, most notably with the great reorganisation in 1974, making long-term comparisons difficult. But the overall pattern is clear. In the late nineteenth century woodland was concentrated in a band running from the south of England into the west Midlands; to the north and east, and to a lesser extent the south and west, there were fewer woods (above, p. 152). By 2000 woodland had increased everywhere and, while the earlier pattern was still visible, it had been disrupted by, in particular, the extensive Forestry Commission plantings in East Anglia and the north (Figure 7.6).

Of course, in many ways all these bald figures are misleading, and the development or establishment of new woodland has not necessarily been beneficial in environmental or, indeed, in any terms. A high proportion has developed, or been planted, at the expense of heaths, moors, fens or other semi-natural habitats of importance for biodiversity. Moreover, the overall increase in woodland area has been accompanied by a reduction in the extent of actively managed, long-established coppice, of prime conservation importance – much of it, once redundant, replanted with conifers. The rate at which coppicing declined in different districts – as a consequence of the spread of coal use and the replacement of many of its main products by manufactured or imported materials – is hard to assess owing to the fact that estimates by different authorities do not always or clearly distinguish between coppice woods *per se*, and those still actively managed. Government figures from 1895 thus suggest that the latter made up only 28 per cent of woodland in England, but as late as 1924 roughly the same percentage was described as 'coppice' or 'coppice with standards' in a Forestry Commission Survey. Indeed, the line between managed and unmanaged coppice is a vague one – in many cases coppices came to be felled at longer and more infrequent intervals, and on an increasingly casual basis, before management was abandoned altogether. The speed of decline also varied from region to region, with coppicing largely disappearing from the north by the end of the nineteenth century, as we have seen, but continuing in southern counties well into the twentieth, especially where active markets for particular kinds of pole existed. In Kent and Sussex the demands of the hop fields kept coppicing alive well into the second half of the twentieth century (Collins 1989). Even in Hertfordshire, the industrial malthouses in the east of the county continued to be fuelled with hornbeam cut from the local woods within living memory. As late as 1970 it was reported that Taylor's of Sawbridgeworth still used 'hornbeam faggots exclusively, of which immense stacks like haystacks are built close by' (Branch Johnson 1970, 33). This said, in most southern and Midland districts coppicing came to an effective end during the inter-war years: typical was the situation on the Eastnor estate in Herefordshire in the early 1930s, where an auction of coppice lots attracted no purchasers at all (Watkins 2014, 182). An increase in the area of woodland thus needs to balanced against profound changes in its character. This said, the figures presented above at least allow us to see more clearly that, insofar as a problem exists, it relates less to the *extent* of woodland in England, than to the ways in which it is used and managed.

The decline in tree health

Worries about the destruction of England's trees and woods, while justified in many ways, are thus to some extent over-played. A far more serious concern, as most readers will be aware, is the way in which our trees appear to be becoming less healthy. This is largely as a consequence of the arrival of a succession of invasive bacteria, insects and fungi, something widely explained as a result of increases in the scale of international trade, climatic change, or both (Brasier 2008). The first of such foreign pathogens to appear was probably oak mildew, first noted in England in 1908 (Mougou *et al.* 2014; Takamatsu *et al.* 2007; Marcais and Desprez-Loustau 2014). Caused by the fungus *Erysiphe alphitoides*, probably of Asian origin, it affects younger leaves and soft shoots in late summer and, while it does not kill affected trees, it can weaken them, leaving them susceptible to other problems. Dutch elm disease, caused by the fungus *Ophiostoma ulmi*, disseminated by the elm bark beetle *Scolytus*, came next, in the 1920s. It may have infected tree populations in England in earlier periods, although many supposed examples of previous outbreaks were probably caused by infestations of the beetle alone, which could be locally serious. A second epidemic, beginning in the 1970s and this time caused by a combination of *Scolytus* and *O. novo-ulmi* (a fungus probably originating in Japan), was far more serious, leading to the effective elimination of elm as a tree (Brasier 1991; Gibbs *et al.* 1994; Clouston and Stansfield 1979). This was particularly devastating given that, as we have seen, elm was in most districts the second or third most common farmland tree (Figure 7.7). In 2000 bleeding canker in horse chestnuts appeared, having been recognised in the USA as early as the 1930s, and by 2007 around half the horse chestnut trees in Britain were exhibiting symptoms. This was soon followed by horse chestnut leaf miner (*Cameraria ohridella*), the larvae of which mine within the leaves of the tree, destroying the tissues. First observed in Macedonia in 1985, this spread steadily through Europe and was first recognised in London in 2002 (Straw and Williams 2013; Tilbury and Evans 2003). In 2005 the oak processionary moth, a major defoliant across Europe, arrived. But most serious of all is the recent appearance of Chalara caused by the fungus *Hymenoscyphus fraxineu* (Cheffings and Lawrence 2014). This causes

7.7 Dead elms in a hedgerow in Bedfordshire, photographed in the 1980s. Almost all the victims of Dutch elm disease have now been taken down, leaving little obvious trace of its terrible impact on the landscape.

severe leaf loss, crown dieback and bark lesions in affected trees, with high mortality rates in young specimens and slower decline in mature examples. The scale of its impact is still unknown, in part because, while first recognised as recently as 2012 – its spores apparently brought by wind from Europe, although imported plant material was clearly a factor in some outbreaks – there are indications that it may have been present in the country for over a decade. A range of other alien pathogens have either appeared briefly in this country (such as sweet chestnut blight, *Cryphonectria parasitica*) or are a potential threat (such as the emerald ash borer, *Agrilus planipennis*).

In addition to these clearly defined illnesses, with identifiable causes, a number of less specific ailments are causing concern among foresters and others, not only in England but across Europe as a whole. The most important is 'oak decline', a syndrome currently thought to have multiple and complex, although debated, causes. Affected trees exhibit progressive thinning of the crown and loss of branches, in some cases leading to the eventual death of the tree. The condition was first described and defined in England in the 1920s and was initially interpreted as a consequence of defoliation by caterpillars of the indigenous oak leaf roller moth (*Tortrix viridana*), combined with the effects of the newly introduced oak mildew fungus. But it was increasingly noticed during the middle and later decades of the century and is now variously attributed to disease and environmental change (Thomas *et al.* 2002) (Figure 7.8). During the 1980s a more extreme form of the syndrome, 'acute

7.8 Oak tree, showing the thinning of the crown typical of the condition known as 'oak decline'.

oak decline', was identified (Denman and Webber 2009). This produces vertical, weeping fissures in the bark and often leads to death within six years (Denman *et al.* 2012 and 2014; Brown *et al.* 2014). The larval galleries of the native beetle *Agrilus biguttatus* are frequently found when infected specimens are felled and it is probable that the insect acts as a vector for a range of bacteria (Moraal and Hilszczanski 2000), although some have suggested that the insect's presence is more a symptom, than a cause, of the disease (Wargo 1996). The rather non-specific character and uncertain causation of oak decline (in its non-acute forms) are paralleled by the condition known as 'ash dieback', recognised well over two decades before the identification of ash chalara in Britain, and first interpreted by some as a consequence of 'acid rain'. It, too, is characterised by severe reduction of the crown and general loss of condition. Again, its causes are debated and its effects can now be confused with those of chalara: it may have masked the first onset of the latter disease.

In attempting to place current concerns about tree health within historical context we ought to begin by noting that while there is little evidence for large-scale epidemic disease prior to the twentieth century, poor health in trees is not in itself new. Even a cursory examination of early texts and documents shows that disease and death have always been normal in trees, as they are in humans. Oak, for example, has always been prone to fungal attacks from *Laetiporus sulphureus* (cuboidal brown rot of heartwood), *Stereum gausapatum* (pipe rot) and *Fistulina hepatica*. The caterpillars of the indigenous micromoth *Tortrix viridana* have long fed on young oak leaves, sometimes causing successive years of defoliation that severely weaken trees. The condition known as 'shake', which affects oak, sweet chestnut and some other timber trees – that is, internal (longitudinal) splitting of the timber, either across or between the growth rings – has for centuries been of particular concern to foresters. Its precise causes continue to be debated, although they appear to be mainly physical rather than biological in character, with trees on well-drained sites, especially on acid soils, being particularly prone (Price 2015; Cinotti 1989; Mather and Saville 1994). Indeed, the symptoms of such long-established native illnesses and pathogens can sometimes be confused with those of recent arrivals. Some of the symptoms of chalara, for example, are comparable with those caused by the common fungal pathogen *Nectria galligena*, the bacterial pathogen *Pseudomonas savastanoi pv fraxini*, the ash bark beetle *Leperisinus varius* or the ash bud moth *Prays fraxinella*. There are also, we might note, signs that some tree pathogens may have become *less* important over time. In particular, the effects of the common cockchafer (*Melolontha melolontha*) on trees, especially young ones, appears to have been much more serious in the eighteenth and nineteenth centuries than today. The beetles often caused widespread defoliation of oaks, as in 1787, when those around Doncaster 'were entirely stripped by them' (Nichols 1795, 31). On one of the plantations of the Duke of Portland in 1788 it was reported that 'the oak and beech leaves are chiefly destroyed by the cockchafer or locust, and the trees appear quite naked'. In some seasons trees were not only extensively defoliated but 'otherwise much injured by these voracious animals' (Donovan 1799, 31), apparently as the larvae ate the roots. Ash trees were widely attacked in this way. Those in one Kentish plantation, 'by having the bark and fibres of their roots preyed on by these destructive insects, made little or no shoots. Some were entirely killed' (*Trans of Soc. For the Encouragement of Arts* 1795, cxcii).

Early writers on farming and forestry seem to have accepted large scale morbidity in trees as something perfectly normal. William Lawson in 1618 pointed to the number of woods 'wherein you shall have, for one lively, thriving tree, four (nay sometimes 24) evil thriving, rotten, and dying trees', with withered tops and falling branches (Lawson 1618, 41). Moses Cook advised that coppices should not be allowed to grow too high, because 'you cannot come to survey your Timber-trees, to see which are decaying … Why should any reasonable Man let his Trees stand in his Woods, or elsewhere, with dead Tops, hollow Trunks, Limbs falling down upon others and spoiling them, dropping upon young Seedlings under it, and killing them?' (Cook 1676, 163). He devoted a short chapter to 'Diseases of Trees', paying particular attention – as did other writers of the time – to the effects of rot, canker and 'worms', and to the importance of keeping caterpillars off trees. The latter is presumably a references to the oak leaf roller moth *Tortrix*, while his widely shared concerns about galls and rots probably reflect the endemic character of *Laetiporus sulphureus* and *Stereum gausapatum*. Many of these problems, he believed, could be remedied by astute surgical pruning. 'If Worms are got between the Bark and the Body of your Tree, they must be cut out'; 'If a Tree be blasted in part, or the whole Head, cut all that is blasted or dead, close off to the quick, and take out all dead Boughs' (Cook 1676, 160–63). Other early writers, such as Mortimer, also refer to the effects of mildew in oak 'and other trees, whose leaves are smooth', as well as to honeydew on oak, maple, hazel and lime. Right through the eighteenth and into the nineteenth century galls, cankers and bleeding remained a perennial concern and the need to prune dead branches from oaks and other timber was universally emphasised, to 'prevent the mortal disease reaching the trunk' (Cobbett 1825, para. 437). There are also clear descriptions of diseases such as 'shake' not only in oak and chestnut but also in elm (Nicol 1799, 159–60; Monteath 1824, 420). References to stag headedness, and to what sound like the symptoms of 'decline', are frequent: 'That dead-topped oaks are very common, cannot be disputed' (Pontey 1805, 130). In 1842 Selby summarised the vast range of threats to trees, discussing the severe damage caused to oaks by *Tortrix viridana* and other insects; the prevalence of canker in larch; and the impact of several insects on elm trees, of which *Scolytus* 'Was considered, some years ago, as by far the most formidable and destructive, as to its ravages in the larva state were attributed the decay and subsequent death of the finest Elms in the vicinity of London, particularly those in St James' and Hyde Parks'. Johns similarly described, at a slightly later date, how 'the ravages committed by this small creature would scarcely be credible' (Johns 1899). But already, in the nineteenth century, the multiple and complex causes of problems such as these were understood, Selby adding how '*Scolytus* is not always, indeed perhaps but seldom, the proximate cause of decay, and that the trees were not attacked by the pregnant females for the purpose of depositing their eggs beneath the bark, until they have become injured and diseased from some other cause' (Selby 1842, 114).

Local documentary sources similarly suggest that ill-health in trees was endemic. The terms used to categorise trees in a survey of Staverton in Northamptonshire in 1835, for example, included 'decayed', 'damaged', 'small and very bad', 'very bad', 'decayed very bad' and 'dead'. In the woods at Geddington Chase in 1760 it was said that 'a great deal of the timber' was decayed (Boughton House archives), while at Mundford in Norfolk in 1805 the surveyor was 'disappointed in the quantity of trees in the Square Plantation, finding such a quantity of dead ones' (NRO MS 13751, 40E3). In the 1760s on the Boughton estate in Northamptonshire 'a great deal' of the timber 'falls very faulty and rotten and makes but very little money', while on another occasion it was said that 14 out of 101 trees inspected in one estate wood were 'so very red and defected that the body will by no means do for shipping' (Boughton House archives). At Prior Royd wood near Sheffield in the late eighteenth century a quarter of the 21,000 wavers, or young successor trees, were reported to be 'very sickly' (Shef ACM/MAPS/Shef/169), while on the Bolton Estate in Wensleydale in 1809 many of the trees at Capplebank were 'much decayed in the middle' (Nrth ZBO 4404). Felling and sales records often refer to dead trees, in spite of the fact that the majority of diseased specimens appear to have been taken down long before they actually died. References to dead ash trees are particularly common, perhaps because, being less valuable as timber, they were less carefully monitored than oak and thus less likely to be felled before problems became acute. The categories of trees sold from the Evans-Lombe estate in Norfolk in 1835, for example, comprised 'Spruce, hornbeam, dead ash, and elms' (NRO HNR 465/3/1). Not surprisingly, given problems with *Scolytus*, 'dead elm' is also often encountered as a specific category in forestry records, as on the Stradsett estate in west Norfolk in 1815 (NRO BL/BG 5/3/4). Trees exhibiting what we would probably now describe as 'oak decline' are often referred to. At Prior Royd wood the 21,000 wavers were accompanied by 206 'reserve' trees that were 'nearly all dead top'd' (Shf ACM/MAPS/Shef/169), while the owners of the Bolton estate in the eighteenth and nineteenth centuries 'generally used the tops of old trees for mine timber, an indication that many of the trees were affected by crown dieback' (Dormor 2002, 222).

In short, poor health in trees was not unusual in the past. It was simply tolerated, with diseased specimens being felled and sold for the best price possible. Nor should we forget that, then as now, trees were also subject to attack from a range of animals. Cook discussed problems caused by deer, hares, rabbits, moles, mice and rooks, and local sources do likewise. At Settrington in Yorkshire in the 1590s it was reported that rabbits from the manorial warren 'do great hurt in the wood by destroying the young springs and they have altogether destroyed one springe called Peeke spring' (King and Harris 1963, 32). All in all, the problems faced by the woodsman were neatly summarised by Hamilton in 1820: 'forest-trees of all descriptions, are subject to such a variety of mischiefs and diseases, even under the most judicious management, that it is to be ranked among the favours of providence that the sustenance of man does not depend upon them' (Hamilton 1820, 446).

It is possible that, even before the rash of epidemics in the twentieth century, the incidence of disease was increasing as forestry practices changed through the later eighteenth and nineteenth centuries. In particular, the way that trees (often of only one or a limited number of species) were planted densely in the new plantations of the eighteenth and nineteenth centuries may have increased morbidity. In 1775 Boutcher referred to the 'many mortal diseases incident to large and crowded plantations' (Boutcher 1775, 247). Plantations were often established on the kinds of poor, acid soil on which trees were particularly susceptible to 'shake': in 1803, after extensive new plantings had been undertaken on upland moors, the agent of the Bolton estate in Yorkshire complained that many trees were affected by 'shakes' (Dormor 2002, 223). By the end of the century Nisbet, in his revision of James Brown's *The Forester*, could note how 'in artificial forests and plantations ... . Trees grow under conditions that are not so favourable to normal healthy development as wild growth in natural woodlands ... . For the plants are not indigenous to the soil, and more or less of one age and are nearly all at equal distances apart from each other' (Nisbet 1894, II, 285). Larch, widely planted in this period, was especially susceptible, but other conifer species, including Scots pine, also exhibited signs of poor health in plantations (Loudon 1838, 2601). All this said, the increased references to and discussions of tree disease that are evident in the course of the nineteenth century probably reflect not so much the deteriorating health of trees and woods but rather the development of new attitudes to forestry – less fatalistic, and more optimistic about what could be achieved through the application of science.

This culminated in the establishment in 1882 of the English Arboricultural Society (later the Royal Forestry Society) and in the growing influence of German forestry practice in England, manifested in particular in the publication of English translations of Herman Furst's *The Protection of Woodlands Against Dangers Arising from Organic and Inorganic Causes* in 1893 (Furst 1893) and of R. Hartig's *Diseases of Trees* in the following year (Hartig 1894). Indeed, there was something of a rash of writings on tree health around the end of the century, including the discussion of pests in H. Marshall Ward's *The Oak*, published in 1892; W.R. Fisher's volume on 'Forest Protection' in Schlich's classic multi-volume forestry text of 1895; and the relevant chapters in John Simpson's *The New Forestry* of 1900 and Percival Maw's *The Practice of Forestry* of 1912 (Ward 1892; Fisher 1895; Simpson 1900; Maw 1912). Of particular interest is Charles Curtis' *The Manifestation of Disease in Forest Trees* of 1892, the first book entirely devoted to the subject of tree health. This describes the kinds of problem alluded to in earlier texts and documents, but with greater clarity. Curtis reviewed the problems associated with particular trees, such as the larch (planting on unsuitable sites, overcrowded plantations leading to infestation by the fungus *Peziza Willkommii* and the development of canker, as well as attacks by the larch bug *Chermes laricis*) (Curtis 1892, 21). He paid special attention to stagheadedness, which he discussed mainly in the context of oak but also of lime. He also listed the various insect pests of trees:

the ash bark beetle (*Hylesinus fraxini*), ash bark scale (*Chionaspis fraxini*), elm bark beetle (*Scolytus destructor*), goat moth (*Cossus ligniperda*), buff tip moth (*Pygera bucephala*), cockchafer (*Melolontha vulgaris*), marble gall fly (*Cynips Kollari*), oak leaf roller moth (*Tortrix viridana*), pine beetle (*Hylurgus piniperda*), pine saw fly (*Lophyrus pini*), pine weevil (*Hylobius abietis*) and spruce gall aphis (*Chermes abietis*). (Curtis 1892, 33)

These caused a host of major and minor problems, including 'the destruction of the leaves of the oak [and] the black fungoid spots upon the leaves of the sycamores and maples'. In most cases, he suggested, insect attack was best dealt with by clear-felling the affected areas of woods or plantations. Curtis singled out the attacks of *Tortrix* as particularly damaging: 'Whole districts and whole woods are sometimes stripped bare of leaves, so as to give an appearance of winter' (Curtis 1892, 41).

Poor health in trees has thus always been with us: what is new is the rising wave of epidemic disease that began in the early twentieth century. According to many scientists, one of its principal causes has been the steady increase in the amounts of wood, timber and plant materials being moved around the world. Yet here again an historical perspective is useful. Even in the Middle Ages much timber was imported, especially from the Baltic, and the scale of imports increased steadily through the post-medieval period. In 1612 a single consignment unloaded at King's Lynn in Norfolk included 900 Norway deals, 1,400 small spars, 436 cant spars, 2,000 stave hoops for casks and 100 two-foot fir timbers, all from North Bergen in Norway. By the eighteenth century significant quantities of timber were also coming from the Americas and by 1905 oak timber was being imported even from Japan (Latham 1957). The scale of imports rose rapidly with the onset of industrialisation, reaching 2.6 million cubic metres by 1851, 5.9 million by 1871, 13.4 million by 1900 and 16.4 by the outbreak of the First World War (Fitzgerald and Grenier 1992, 18). Following the war there was a decline, but this was soon followed by a second peak, of 16.2 million, in 1936 (Fitzgerald and Grenier 1992, 144). Since then the volume of imports has followed an erratic course, but currently stands at around 6.3 million cubic metres of sawn wood per annum, in addition to 3.2 million cubic metres of wood-based panels and 9.7 million cubic metres of wood pellets (Fitzgerald and Grenier 1992, 144; http://www.forestry.gov.uk/forestry/infd-9nrkgg).

Live tree materials were also being moved long distances from an early date. This initially involved fruit trees and ornamentals for gardens – in November 1696 the Norfolk landowner Richard Godfrey lamented that the weather was preventing the delivery of the fruit trees he had ordered from Holland (NRO Y/C 36/15/18) – but soon forest trees were being imported. In 1700, for example, John Bridges of Barton Seagrave in Northamptonshire planted '500 limes from Holland' (Morton 1712, 486). By the end of the nineteenth century large quantities of live forest trees were being brought from abroad on a commercial scale, especially conifer seedlings from France and Germany, and in 1910 the noted forester P.T. Maw urged the governing council of the Royal Forestry Society to:

> Put forth every effort in their power to induce the Board of Agriculture to … prohibit the importation of seedlings raised abroad … a most dangerous practice and one calculated to destroy the comparative immunity from fungus attacks which at present our young plantations enjoy. I am quite aware that many nurseries are stocked with continental seedlings but I can assure those who adopt this course that they are only 'buying trouble'.
> (Maw 1910)

Given that imports of both wood and plant materials were long-established, however, it was probably less an increase in their scale but rather one in the speed with which they now took place that led to the appearance of oak mildew in 1908 and of Dutch elm disease a few years later. By 1870 a number of inventions, including the screw propeller, the compound engine and the triple-expansion engine, made the shipping of bulk cargoes (as opposed to passengers) by steam, rather than wind, economically feasible, and goods thus travelled much faster (Carlton 2012). Indeed, by this time European diseases were spreading to other parts of the world. Beech bark disease, common in Europe, arrived in Nova Scotia in 1890; a European phytopthera appeared in Maine in 1930 (Ehrlich 1932; Chester 1930).

While the increasing scale and speed of transport systems steadily increased the threat from invasive pathogens through the twentieth century, other factors   less frequently noted by plant scientists – may have contributed to a more general decline in tree health. The most important was the gradual increase in the number of old trees in the countryside. As we have seen, in the period before the mid-nineteenth century most timber trees, both on farmland and growing in woods, were felled before they were 50 or 60 years old. From the 1850s average age rose steadily, and from the early twentieth century it increased still further, as the intensity with which farmland trees were managed gradually declined: the numbers of new trees being planted dwindled, and existing timber, if it survived opportunistic fellings, often grew to an advanced age. The reasons for these latter developments were complex. At the start of the twentieth century most of the land in England was in the hands of large landed estates that operated integrated forestry programmes, embracing both woods and plantations *and* timber growing on farmland. Many of these large properties were broken up in the inter-war years and most of their land passed into the hands of 'farmers who know little and care less for forestry', and whose main concern was the growing of crops and the raising of livestock (Butcher 1941, 361). Even where large estates survived, forestry practices changed. The scale of imports, and consequent low prices, made it difficult to justify, in financial terms, the felling and transportation of single trees growing in isolated locations, and as a result commercial forestry was concentrated more and more in woods and plantations, and farmland trees were left unmanaged. And as farming became more industrialised in character in the course of the twentieth century, many of the features or equipment formerly constructed with timber grown on estate land – fencing, gateposts, minor buildings – were now supplied by manufacturers.

In addition to all these essentially economic and silvicultural factors, there were important changes in social attitudes towards trees. While large landowners had always appreciated the individual beauty of trees, such sentiments were, from the late eighteenth century, further fuelled by the 'picturesque' ideas of men such as William Gilpin, Richard Payne Knight and Uvedale Price. More of the trees on their properties were regarded in aesthetic rather than simple economic terms, although initially in limited locations, in parks and on home farms. Such attitudes gathered strength over time, however, and found support in other quarters. The late nineteenth and the early twentieth centuries saw the establishment of a number of organisations dedicated to the conservation of rural landscapes, open spaces and wildlife. These included the Commons Preservation Society (1865), the Society for the Protection of Birds, later the RSPB (1889), the National Trust (1895), the Society for the Promotion of Nature Reserves (1912) and the Council for the Protection of Rural England (1926), the aim of which was the preservation of 'all things of true value and beauty' (Evans 1992; Cowell 2002; Waterson 1994; Sheail 2002). Arguably, the appearance of these bodies signalled a growing divorce of the majority of the population from the realities of rural life. By 1851 half the population of England already lived in towns and cities, but by 1911 this had risen to 80 per cent (Sheail 2002, 12). Moreover, in the first half of the twentieth century, owing to significant changes in the distribution of wealth and improvements in transport, an urban-based and increasingly mobile group visited the countryside on a larger and larger scale. As the century wore on they increasingly settled in it, or in suburbs on its margins, and began to take an active interest in its conservation. The idea that the countryside was essentially 'natural', which had been developing (alongside urbanisation and industrialisation) since the eighteenth century, now triumphed. Felling prominent hedgerow trees gradually came to be regarded, even by many landowners, not as a normal part of land management but rather as a desecration. Such ideas were manifested with particular clarity, somewhat paradoxically, where countryside was being lost to urban or suburban development. It was proudly claimed that Letchworth Garden City, established in north Hertfordshire in 1902, had been built on virgin farmland without the loss of a single tree (Rowe and Williamson 2013, 274). In 1912 sales particulars for a suburban development in the south of the same county emphasised that 'It is desired to preserve the rural characteristics of the locality as much as possible, and with that object in view the natural hedges and as many of the trees will be retained as is consistent with convenient development' (Bushey [Herts] Museum archive).

By the time of the Second World War the idea – long promulgated by land-use planners such as Patrick Abercrombie and campaigners such as Clough Williams Ellis (Abercrombie 1943; Williams-Ellis 1928) – that state intervention was required to preserve the rural landscape from large-scale development was widely accepted, culminating in the Town and Country Planning Act of 1947 (Rowley 2006, 112–15). As well as introducing, for the first time, workable systems of spatial planning, this also established Tree Preservation Orders, which allowed specimens deemed to be of particular value to be preserved from felling. Although largely applied in urban areas, TPOs represented the triumph of the new attitude to trees – as objects of the natural world to be preserved, rather than as economic objects to be husbanded and exploited.

The increasing age of farmland trees that arose from these complex causes was manifested, in particular, in the growing incidence of 'stag headedness', or dieback. Photographs of the countryside dating from the late nineteenth century show, by modern standards, remarkably few stag-headed trees. Those taken in the post-war period, in contrast, show far more. By the 1950s and 1960s the ageing character of trees in the countryside was becoming a matter of real concern. The *Report of the Committee on Hedgerow and Farm Timber* (Forestry Commission 1955) emphasised how trees felled or dying were not being replaced by new planting or natural regeneration on a sufficient scale to maintain numbers. Around a third of hedgerow trees had girths in excess of 1.5 metres, suggesting an age of at least 60 years – the age by which, a century earlier, most would have been felled (Forestry Commission 1955). An unpublished Forestry Commission Survey of 1972 suggests that 17 per cent of farmland trees were stag-headed and 1 per cent were dead in Northamptonshire, 15 per cent and 3 per cent in Hertfordshire, and no less than 45 per cent and 6 per cent in Norfolk (private archive). The fact that Norfolk, a largely arable county, displayed the highest numbers of ill-looking trees suggests that the increasing intensity of agriculture in the post-war decades was also a major factor in declining tree health. The Forestry Commission survey of 1972 certainly acknowledged the importance of field drainage, ploughing too close to hedges, pollution and stubble burning as threats to farmland trees, although it suggested 'general old age' as the main reason for their deteriorating condition.

It might be objected that the countryside has always been filled with large numbers of senescent trees because, before the middle decades of the nineteenth century, a small number of relatively young timber trees had been accompanied in many districts by a much larger population of ancient pollards, retained because they produced a reasonable crop of poles for centuries. But, as we have seen, while pollards generally attained a greater age than timber trees they, too, were usually replaced with younger specimens as their productivity declined and few, perhaps, survived for more than two centuries (above, p. 51). In addition, we should perhaps note that actively managed pollards were anyway maintained, in effect, in a state of permanent juvenescence (Read 2008, 251). In Lennon's words:

> Because the tree is regularly being cut back and the crown is constantly having to reform, pollarding can delay the emergence of the tree from the formative growth period. Where trees are continually pollarded the ring width will remain trapped in the formative cycle. This can extend the natural lifespan of the tree significantly. (Lennon 2009, 173)

As pollards ceased to be lopped, however, they would begin to age in a normal fashion.

For a variety of reasons, therefore, the modern countryside now contains much larger numbers of senescent trees than would have been the case a century or so ago, and it is within this context that we need to consider some of the current concerns about tree health, and especially the significance of those vaguer conditions, oak decline and ash dieback, which principally affect specimens a century or more in age. In historical terms, these are over-mature trees, and it is possible that these conditions are, to an extent, simply symptoms of normal ageing, transformed into a 'disease' by modern and unrealistic expectations of perpetual arboreal health. More importantly, when tree populations were rigorously managed few specimens would have exhibited such symptoms for long, for the simple reason that they were taken as a sign that a tree was ready to be felled. A tree in 'decline' was one whose useful growth was over and it was sensible to take it down before its timber value was reduced by the onset of decay, and get another growing in its place. As Moses Cook put it in 1676:

> When a Tree is at its full Growth, there are several signs of its decay, which give you warning to fell it before it can be quite decay'd; as in an Oak, when the top boughs begin to die, then it begins to decay; in an Elm or Ash, if their Head dies, or if you see wet at any great Knot, which you may know by the side of the Tree being discolour'd below that place before it grows hollow … these are certain Signs the Tree begins to decay; but before it decays much, down with it, and hinder not your self. (Cook 1676, 171)

Although 'oak decline' was formally named and characterised only in the 1920s, trees exhibiting the appropriate symptoms are often referred to in early texts, and on an apparently increasing scale from the nineteenth century. Curtis in 1892 described how 'dead upper branches or "stag-horn top," as it is usually called, is often met with … . The manifestation needs but little remark, for it is apparent to all. The top branches die, the yearly growth is meagre, and the whole tree presents an enfeebled condition' (Curtis 1892, 25). It is noteworthy, however, that he drew attention to the prevalence of the condition not on farmland or in woods but 'on lawns and pleasure grounds … and park lands' – that is, in locations where many trees were already, by the late nineteenth century, being retained beyond economic maturity. The spread of the condition more widely, in other words, may simply reflect a decline in intensive management and an increase in the proportion of mature and over-mature trees in the countryside as a whole from the late nineteenth century.

Other changes to the rural environment over the last century and a half may have contributed to an increasing level of arboreal ill-health, and should be briefly noted. In many districts, water tables and soil moisture levels are lower than in the nineteenth or early twentieth centuries – seasonally at least – owing to the increased scale of land drainage and water abstraction and changing patterns of cultivation, with a shift to late summer cultivations and continuous courses of crops. This development has often been highlighted. Less attention has been paid to the marked increases in soil nitrogen levels that have occurred since the eighteenth century. The adoption of crops such as turnips and clover, the cornerstone of the agricultural revolution of the later eighteenth and early nineteenth centuries, was specifically intended to increase stocking levels, and thus manure supplies, by allowing more livestock to be over-wintered, while clover also served to fix nitrogen in the soil directly from the atmosphere (Beckett 1990, 11–18). From the 1820s the use of manufactured oil cake as feed rose steeply, further raising livestock numbers and thus the amount of dung applied to the land: national consumption increased from around 24,000 tons per annum in 1825 to 160,000 in 1870 (Thompson 1968, 73–4). Around the same time natural fertilisers began to be imported from abroad, in the form of guano from South America, while bone dust began to be applied to arable fields from the 1830s and then, from the 1840s, a range of manufactured fertilisers, the use of which has continued with varying degrees of intensity until today. Serious eutrophication, resulting from increased nitrogen in water draining in from farmland, was noted at places such as the Norfolk Broads as early as the 1850s (George 1992, 105–12). Increased nitrogen levels may have implications for tree health given that accumulating research suggests that inorganic fertilisers can suppress the development of mycorrhizal fungi, on which tree health depends (e.g. Ryden *et al.* 2003).

A more significant contributory factor may be the marked rise in the quantity of dead wood in the environment over the last century or so. Before the middle of the nineteenth century every possible source of firing available in the countryside appears to have been exploited, especially by the poor, and there were regular prosecutions not only for the theft of firewood but also for 'hedge-breaking', the illicit removal of wood from hedges, although some of these relate to social protest and enclosure disputes rather than to the actual gathering of fuel (Blomley 2007; McDonagh 2009). In such circumstances, fallen branches and even twigs would have been rapidly removed from the environment on a massive scale. Modern conservation practice rightly encourages the preservation of dead wood, including the practice – unknown in the past – of allowing completely dead trees to remain standing for long periods. The steady accumulation of large quantities of this material in woodland and in hedges is, in historical terms, a relatively recent development and one which may have implications for tree pathogens. The native buprestid beetle *Agrilus biguttatus*, thought by many to be a factor in acute oak 'decline', was until recently considered a 'red book' species, to be encouraged by the retention of fallen wood. An earlier generation of foresters was clear about the potential threat posed by large accumulations of decaying wood: 'At the risk of repetition I would impress upon all foresters the necessity of cleaning up after every fall of timber, and the total destruction by fire of all dead organic matter' (Curtis 1892, 46).

It is unlikely that any of the changes discussed over the last few pages, including the marked rise in the average age of trees, can have been significant influences on the scale or incidence of the various epidemics, caused by invasive organisms, that have appeared in this country over the last half century. Dutch elm disease and ash chalara, in particular, appear to infect young specimens at least as much as senescent ones. Indeed, the current impact of the latter disease has mainly been on immature trees. This said, it might be noted that if trees displaying signs of such illnesses were immediately cut down this might serve to retard the spread of some pathogens, and such rigorously managed tree populations would at the very least *appear* to be healthier than under-managed or wild ones. 'Quarantine felling' – that is, the immediate and systematic removal of diseased trees together with a 'buffer' of their still healthy but vulnerable neighbours from woods and plantations – is less regularly practised than in earlier periods, when foresters such as Curtis recommended it as the best way of dealing with problems such as *Hylesinus fraxini*, *Chionaspis fraxini* and *Scolytus*. The message from history may be not simply that disease is a natural condition of trees but that the most unnatural and most rigorously managed tree populations are also the most healthy ones.

CHAPTER EIGHT

Conclusion: past and future treescapes

Over the course of this book we have traced some of the complex ways in which English trees and woods have developed in the period since the end of the sixteenth century. We have emphasised – perhaps over-emphasised – the essentially unnatural character of our tree populations and the manner in which they have been structured by economic and social, as much as by environmental, influences. Until relatively recently, most trees existed in particular places because they were deliberately planted or, if self-seeding, were permitted to grow there. The balance of species found in different locations was thus in large measure a consequence of human choice or management systems. The dominance of woods and farmland by oak, elm and ash trees reflected the many ways in which the wood and timber of these species could be used and their ability to thrive in a wide range of conditions. Other kinds of tree were less useful, less productive or more demanding – harder to grow or process. Nevertheless, other species might be encouraged in particular locations or circumstances. Beech and hornbeam, for example, were favoured in some districts in wood-pastures; willows were numerous on alluvial wetlands. Soil conditions and climatic factors unquestionably had an influence on trees and woods, but in the last resort human choice was key. It may even have had a determining influence on the composition of the understorey in many coppiced woods as, over the centuries, species of little utility were systematically removed and replaced by those producing wood of greater value.

Trees and woods were also managed in the past with an intensity that we would today find surprising, perhaps unsightly, and in ways that were directly related to economic circumstances – hence the phenomenal density of pollards before the nineteenth century in districts that lacked other fuel sources, the early age at which most timber trees were felled and the varying length of coppicing cycles in different regions. Wood and timber were of central importance and in high demand not only in the pre-industrial but also in the industrialising world. Such intensive management of the environment was not, of course, a particular feature of trees and woods. Most if not all of the 'semi-natural habitats' which exist and existed in England, as in the rest of Britain and Europe, were created and sustained by regular exploitation. Heaths, for example, were very intensively grazed, both by sheep and rabbits, and often in a manner that served to systematically deplete them of nutrients, with the sheep being removed by night to folds on the arable fields, where they dunged the land. Gorse, broom and heather were regularly removed on a large scale for fuel, the digging of the latter serving to add to the ground disturbance created by rabbits and by the widespread excavation of sand and gravel. It was all these activities that prevented heathlands from reverting to woodland, and which gave them their particular flora and fauna. Meadows, in contrast – like coppiced woods – received most of their character from the way in which they lay ungrazed for much of the time. They were closed to livestock during the late spring and summer, allowing plants intolerant of grazing and trampling to flourish, flower and set seed, many of them tall, bulky species such as meadowsweet (*Filipendula ulmaria*), globeflower (*Trollius europaeus*) or oxeye daisy (*Chrysanthemum leucanthemum*) (Fuller *et al.* 2017). In short, intensive management was not and is not inimical to wildlife or to biodiversity, and what we usually think of as 'nature' was largely the consequence of particular forms of past land use: obvious points to many readers, no doubt, but not always, perhaps, to the more enthusiastic advocates of 'rewilding' (above, pp. 11–12).

The character of trees and woods, always no doubt changing in response to economic and social developments, was more radically transformed from the end of the eighteenth century by changes in fuel economies and technologies, increasing levels of long-distance trade, new patterns of ownership and novel systems of land management. It is these developments, as much as the agricultural intensification that devastated the countryside during the post-war years, that have shaped the present condition and character of our trees and woods. The last two centuries have witnessed the effective disappearance of wood-pastures, the steady expansion of plantation forestry, the gradual abandonment of coppice management and an inexorable decline in the numbers of farmland trees. Yet, as we have also suggested, recent arboreal history has not been a tale of unremitting gloom. Although traditional woods have declined – replanted with conifers or, at best, left to become neglected and overgrown – the area of woodland *per se* has expanded steadily in the course of the last century, and in many regions now covers a greater area than in *c.*1600. Moreover, on some measures numbers of farmland trees have also witnessed a significant recovery, although over a more recent period, since the 1970s.

How we evaluate these apparently more benign developments depends very much on our interests and perspective; and here we should be wary of the systemic pessimism that structures some approaches to nature conservation. We value the habitats we have, especially those deemed long-established, with the various life forms they sustain; and we rightly fight to preserve them in an age of frighteningly rapid change. But this in turn means that we sometimes fail to acknowledge the benefits provided by the environments that replace them. Even the impact of industrial development and urbanisation is seldom universally negative. 'If an area of average farm land is developed into a housing estate, the fauna and flora will certainly be different but it would be unwise to assume that it would eventually be any less rich or interesting' (Davis 1976, 282–4). Indeed, one study of a fairly ordinary suburban garden in Leicester, carried out over a period of three decades from the 1970s, recorded no fewer than 2,673 different species of flora and fauna, including butterflies, moths, beetles, hoverflies, mammals and birds: represented were 54 per cent of Britain's ladybird species, 23 per cent of its bees, 19 per cent of its sawflies, 48 per cent of its harvestmen and 15 per cent of its centipedes (Owen 2010). This is far more than most equivalent areas of farmland could muster.

Similar considerations arguably apply to woodland. It is true that the widespread destruction and replanting of ancient woods in the middle decades of the twentieth century led to an appalling loss of biological diversity, but the establishment of new commercial plantations more widely in the period since the 1920s, including those composed entirely of alien conifers, was not in itself an environmental catastrophe. While it certainly led to a marked decline in many of the distinctive plants and invertebrates of the moors and heaths in the planted areas, recent studies have concluded that 'stands of non-native conifer species appear to provide suitable habitat for a wide range of native flora and fauna and should be viewed as making a positive contribution to biodiversity conservation in the UK', especially if managed appropriately (Humphrey *et al.* 2003). Many bird species have certainly benefited from the new woods, such as the green woodpecker and the siskin. The massive expansion in the numbers of red and fallow deer in England since the 1950s is also, in part, a consequence of the new plantations, which provide the large tracts of cover they require, although other factors, including the less disturbed character of the rural landscape as a consequence of the mechanisation of farming, have also contributed. It is noteworthy that many conservationists, rather than celebrating this marked recovery in the fortunes of our largest and most striking wild mammals, see it essentially as a *problem*, because of the damage it has wrought on other species, such as the characteristic plants of ancient woodland (Dolman and Waeber 2008). Conservation judgements are more subjective than we often recognise and biased towards the defence of the status quo, even against a return to aspects of earlier conditions.

The natural regeneration of woodland over surviving areas of derelict heaths and other common land – something which has contributed significantly to the overall increase in woodland area over the past century – is also often viewed in largely negative terms. Heaths in particular are now a rare resource, hence the amount of money and time rightly spent clearing surviving examples of scrub to allow characteristic plants and birds such as the woodlark or the stone curlew to flourish. But heathland is largely if not entirely an unnatural, anthropogenic environment, created and sustained – as just noted – by forms of land use that have now been rendered redundant by economic and technological change. The secondary woodland that has replaced it in many places is not necessarily an impoverished or worthless habitat, and is arguably a more 'natural' one. Birch, the principal element in early stages of succession, was rather a rare tree in lowland areas before the twentieth century, and the massive increase in its numbers is in some ways to be welcomed. Birch trees can support over 330 different species of insect, and woods of birch have high conservation value: light and delicate foliage allows a rich ground flora to develop, in contrast to that often now found within the dense and gloomy stands of neglected traditional coppice. Left to their own devices such woods gradually evolve into ones dominated by species such as oak, potentially of high conservation value on any measure. The various kinds of new woodland that have appeared across England over the last 100 years, in other words, should perhaps be regarded in a more positive light than they often are, and certainly not as an unremitting disaster. And this, of course, is without discussing other benefits – in terms of recreation, for example – that woodland brings to the wider community. As Watkins has emphasised, 'Woods do not have to be exciting or contain rare species to be interesting, and all have distinct histories and unknowable futures' (Watkins 2014, 271).

Evaluating the value of the non-woodland trees that have been added to the landscape over the last four decades is more difficult. As we have seen, the increase in the numbers of farmland trees growing in 'groups' and 'linear features' has almost everywhere been accompanied not only by a continuing decline in free-standing specimens but also by a marked concentration of trees in certain limited areas within the landscape owing to the increase in the size of fields – something especially marked in eastern arable districts. Whatever the absolute numbers of farmland trees, their distribution is now very different from that which pertained in the period before the middle of the twentieth century. This is significant for a number of reasons. A landscape scattered fairly evenly with trees provides habitats for large numbers of mammal and invertebrates, and thus for birds; individual trees also provide important song-posts for species such as whitethroat, wren and robin. Not surprisingly, hedges with trees support a significantly more diverse avifauna than those without, and in a number of ways the two cannot be separated: both have declined together. In combination, they provide the corridors for wildlife that are vital in intensively farmed countryside, linking wooded areas and other uncultivated pockets of land used by a wide range of invertebrates and small mammals, as well as by birds. Hedges provide a prime habitat in their own right, of course, for bird species such as dunnock, yellowhammer and common whitethroat, for mammals such as voles and mice and for a vast range of invertebrates. Their importance for biodiversity in the past was particularly great because regular management provided a rapid succession of habitat structures as the hedge recovered from coppicing or laying. Current management by frequent and relatively unintensive trimming and flailing, in contrast, seldom provides the same range of niches, and the landscape of wide fields, few hedges and fewer free-standing trees that has emerged over large areas of eastern England is less conducive to wildlife. Where hedges have been allowed to grow unmanaged into lines of tall trees, moreover, they may afford good feeding grounds for woodland birds such as redstarts, tits and woodpeckers, but provide a poor habitat for birds characteristic of open farmland (Brown and Fisher 2009; Fischer *et al.* 2010; Forman and Baudry 1984; Peterken and Allison 1989).

But there is another aspect to all this. The English landscape, as we have noted on a number of occasions, has always been changing. But for centuries most of the countryside

8.1 John Constable, *A Lane Near Dedham Vale* (1802). The paintings by Constable and other eighteenth- and nineteenth-century landscape painters have done much to shape our ideas of the 'typical' English countryside in terms of hedges and hedgerow trees.

in lowland districts has comprised fields enclosed by relatively low hedges well studded with individual free-standing trees. This is the kind of landscape captured, for example, in John Constable's paintings, or in those by members of the early nineteenth-century Norwich School (Figure 8.1). Leaving aside issues of nature conservation and biodiversity, the replacement of free-standing trees, distinct from the hedges in which they grow, by 'linear features', certainly creates a landscape subtly but decisively different from what is usually thought of as the 'traditional' English scene. Of course, as we have made clear, that landscape was itself an artefact, created by specific social, economic and technological circumstances, and early nineteenth-century visual representations – by failing to portray such 'ugly' features as lines of newly cut pollards – often present a misleading picture of a rural idyll. But such images and models matter, providing as they do a sense of place and continuity in a rapidly changing world. In short, raw figures presented by the Forestry Commission and other bodies showing the apparent recovery in farmland tree numbers are not quite the unqualified success story that they may appear, in terms of either nature conservation or cultural value. There is, perhaps, an important distinction to be made here between the recorded increase in the numbers of non-woodland trees and that which has unquestionably occurred in the area of woodland.

The way we tend to view farmland trees and woods today as part of a separate, 'natural' world reflects their current, often unmanaged or minimally managed condition, as well as the general divorce of an urban population from the practicalities of rural land management. We probably don't like to think of our ancient woods and hedges as, in essence, redundant factories for the production of wood and timber. We prefer to believe they are fragments of, and a direct link with, the primeval forests, places in which we can find a refuge from the horrors and uncertainties of the modern world. But the rigorous management that pertained before the mid-nineteenth century may, to an extent, have helped to ensure the general good health of trees, for diseased specimens were rapidly felled and few trees reached an advanced age where they would become more susceptible to pathogens. The 'wildwood', in contrast, may have been chock full of diseased and dying specimens. The suggestion that in the past tree populations were younger and more vigorous than those of today might appear to be contradicted by the existence of 'veteran' specimens, particularly oaks, in the landscape: ancient trees of particular importance to biodiversity because of their role as hosts for saproxylic insects, lichens and other organisms which depend on them, and valued by many as a direct and living link with ancient times. It is sometimes assumed that they are survivors from a much larger population of ancient trees, drastically thinned by the ravages of the twentieth century, but, as we have noted, they have probably always been few in number and concentrated in particular areas, such as private parks, where wealthy owners could afford to indulge the same affection for old trees that we share today: certainly, most examples can now be found in such locations (Figure 8.2). Elsewhere, in a countryside whose management was dominated by economic considerations, most pollards were probably removed as they aged and became less productive.

Our particular interest in, and affection for, very old specimens is perhaps related to an important feature of our attitude to trees more generally: an assumption, usually unspoken, that they *should* live to a very great age, and that death and disease are in some way unusual, even unnatural. As the previous chapter should have made clear, trees left to their own devices were always prone to attacks by bacteria, insects and other pests. It was the job of people – carried out very effectively in the pre-industrial countryside – to maintain unnatural levels of health through the early removal of diseased specimens, the quarantine felling of diseased stands and the more general policy of cutting trees down before they came anywhere near old age. Some trees always avoided such treatment, *and* failed to succumb to fatal infection, eventually reaching advanced senescence: but they were few in number.

None of this detracts from the very real threats we now face from successive waves of invasive pests – bacteria, fungi or insects – that are, like the spread of invasive species more generally, a consequence of an increasingly globalised world and more

8.2 A medieval oak pollard, with a circumference of over nine metres, preserved in parkland at Houghton in Norfolk. Most of England's most ancient trees are to be found in parks.

specifically of the increased speeds with which wood, timber and live plant materials can be transported vast distances. Here historical research brings no good news, instead confirming that this is indeed a new development. Although trees have always suffered from illness and disease there is no evidence – at least during the period since *c*.1600 – for large-scale epidemics before the early twentieth century. To some extent, it is true, the scale of present threats may be exaggerated, not least by those whose research funding depends on racking up the anxieties of the public and the government. At the time of writing it does not appear likely that the initial predictions of some scientists that ash chalara might lead to the extinction of ash as a species in Europe will be fulfilled. But the example of Dutch elm disease casts a long shadow, is always with us: and what chalara spares, emerald ash borer, heading fast across Europe towards the ash trees near you, may finally destroy. Given that it seems impossible to turn back the clock and reduce the rate and scale of trade and movement, invasive threats will continue to affect us.

In the face of such problems, does an historical perspective provide any solace or suggest any policies of mitigation? Perhaps. Dutch elm disease was the appalling disaster that it was, and ash chalara the disaster it may yet turn out to be, because elm and ash were or are extremely common trees. If we were to experience a serious epidemic in wild service, in contrast, few people – other than a handful of naturalists and tree scientists – would notice, and the disease itself would move slowly, from one scattered specimen to another. A major problem is thus that our landscape, and especially our farmland, is dominated by a relatively small number of tree species. It is true that recent developments have to an extent changed this: conservation and amenity planting have increased the representation of trees such as lime, hornbeam and maple; the development of some hedges into lines of tall trees has massively increased the importance of the latter species in many areas; while the regeneration of woodland on ungrazed derelict and marginal land has significantly increased the amounts of birch in the landscape. Nevertheless, oak, ash and maple still make up over 80 per cent of our farmland trees. We urgently need to diversify, to ensure future resilience, and here a recognition of the essential artificiality of rural tree populations gives us more freedom in our choices of what we should plant in order to achieve this. We need no longer plant exclusively or primarily with a harvest of wood or timber in mind, and this opens up a wealth of opportunities.

Some authorities argue that we should consider supplementing our palette of native trees with a range of southern European varieties, such as downy oak (*Quercus pubescens*), not merely to increase diversity but also to guard against the likely effects of climate change. While we acknowledge that this may have a place in future policy, we would urge some caution here. Most of our indigenous species, including ash, beech, birch, oak, small-leafed lime, hornbeam, black poplar and aspen, in fact have distributions that continue far south into Europe, and so can evidently tolerate significantly warmer conditions. Less important, perhaps, than the addition of new species is the more extensive use of those that are already here, whether native species or long-established naturalised ones. We have many to choose from. While there are good arguments for continuing the present reliance on oak – and ash, if and when strains resistant to chalara are identified – and for continuing to encourage the current expansion in the numbers of maple, many other species might be planted in woods, copses and in particular in hedges. Small-leafed lime, the beautiful tree which once dominated vast tracts of the landscape, is an obvious choice; black poplar is another, although its large, spreading and fast-growing character might necessitate some thought about appropriate sites. Hornbeam would also be a magnificent addition to the landscape, in southern districts especially. Beech, holly, alder, willows, cherry, rowan, crab, wild pear, whitebeam, wild service, aspen, Scots pine – even sweet chestnut and sycamore, much under-rated in terms of biodiversity – could all be more extensively established, the latter perhaps especially in northern England, where it is already well represented as a farmland tree in many areas (Figures 8.3 and 8.4).

All this said, caution should be exercised over what precisely should be planted where, and not only because some of these species have particular requirements in terms of soils, drainage and the like. Planting an indiscriminate 'conservation mix' in all areas would tend to further erode the sense of regional distinctiveness and local character so important in the English countryside, but which is already much reduced

8.3 A wild service tree at Maulden in Bedfordshire. A rare tree, wild service would make a magnificent addition to the hedgerows of England.

8.4 Wild pear growing in a hedge near Attleborough in Norfolk. *Pyrus communis*, probably a naturalised species, has considerable benefits for wildlife: it is another tree which could be more widely planted on farmland.

by insensitive additions to the built environment and by the widespread simplification – prairification – of the farming landscape. Instead, we should be guided to a large extent by the distributions of what we have described as the 'minority' trees of farmland, for in the past these combined with oak, elm and ash to create distinctive 'treescapes' that could be subtly intensified and accentuated in the future. In the county of Hertfordshire, for example, there were long-established and significant differences between the west of the county – the Chiltern dipslope, with soils largely formed in clay-with-flints and outwash gravels – and the east, with soils mainly formed in boulder clay. Before the nineteenth century oak was the dominant tree in the hedges of the west, accompanied by smaller amounts of ash and elm; but, in addition, cherry was regularly found, alongside smaller quantities of apple, aspen and beech, and with scattered hornbeam on the poorer soils (Figure 8.5). In the east of the county beech and aspen were virtually unknown: instead, in addition to oak, ash and elm, maple and hornbeam were present on farmland, with small quantities of black poplar on damper sites. Moving east, into Essex, these were joined by small amounts of small-leafed lime. Replicating, restoring and accentuating such patterns would ensure that a measure of regional diversity could be perpetuated into the future, providing a 'sense of place' and a measure of historical continuity, which might be lost if new species from abroad, or some fairly standardised and repetitive collection of indigenous species, were to be widely established. In addition, such 'minority' trees are 'tried and true' and likely to succeed in the localities in question. But we could also be bolder. In the area around London attempts might be made to recreate the great wood-pastures of hornbeam, lost from the landscape only relatively recently; in Midland districts, the willows that existed in vast numbers in the old 'champion' landscape might be restored.

Although history cannot tell us what to plant in particular locations it can help us make choices, and it can give us the confidence to plant with a wider palette, safe in the knowledge that past tree populations were in many ways highly artificial, structured by economic and agrarian considerations, and that we now plant with different aims. This said, we should recognise that some of the problems we currently face may not be helped by a complete emphasis on conservation, amenity and aesthetics. It is hard to escape the conclusion that some of the current troubles with farmland trees derive from the fact that we have moved away from the kinds of rigorous management, rooted in commercial considerations, that were normal in the past. Trees have perhaps always had important non-economic roles, and old trees may always have been given a particular value. As Watkins has noted, 'Trees and woods often outlive humans and provide a semblance of order, continuity, and security' (Watkins 2014, 10). But in a fast-changing world preservation and conservation have gained the upper hand over other considerations in many contexts. Urgent research is required into the extent to which the increasing numbers of old trees in the countryside, the general failure to remove diseased examples rapidly, the decline in the practice of quarantine felling and – perhaps – the steady accumulations of dead and

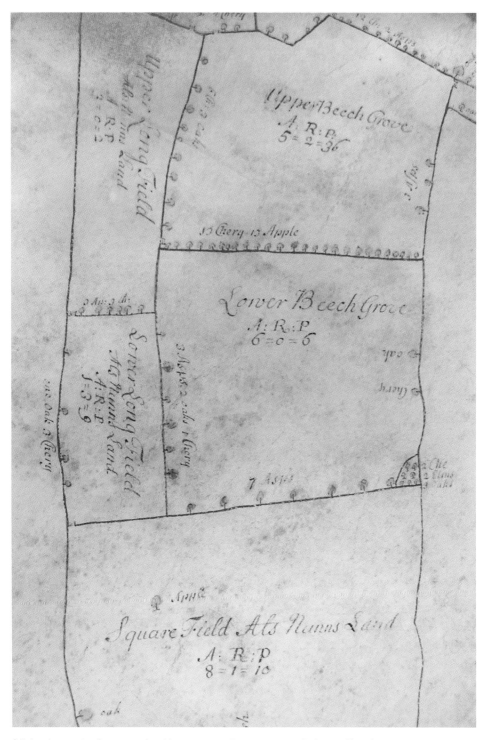

8.5 Another section from an undated late seventeenth-century map of a farm at Flaunden in west Hertfordshire, showing hedges full of cherry, apple and aspen.

decaying wood in woods and hedges may be bringing as many problems as benefits for wildlife, or at least for trees. As we noted earlier, it is possible that the most intensively managed tree populations are the most healthy. But these are issues for natural scientists, rather than historians, to address.

Bibliography

Abercrombie, P. (1943) *Town and Country Planning*. London: Home University Library.

Adams, B. (ed.) (1997) *Lifestyle and culture in Hertford … Wills and inventories for the parishes of All Saints and St Andrew, 1660–1725*. Hertford: Hertfordshire Record Society.

Addington, S. (1978) 'The hedgerows of Tasburgh', *Norfolk Archaeology* 37, 70–83.

Albert, W. (1972) *The Turnpike Road System in England 1663–1840*. Cambridge: Cambridge University Press.

Armstrong, M. (1781) *History and Antiquities of the County of Norfolk*, 10 volumes, Norwich.

Austin, P. (1995) 'Hatfield Great Wood and its enclosure', *Hertfordshire's Past* 36, 2–7.

Austin, P. (1996) 'The leasing of Lord Burghley's Hoddesdon woods in 1595. An insight into woodmanship', *Hertfordshire's Past* 41, 11–21.

Austin, P. (2001) 'Traditional woodland management. The woods of the Manor of Digswell 1776–1843', *Hertfordshire's Past* 51, 2–12.

Austin, P. (2008) 'Traditional woodland management: the timber trees of Panshanger in 1719', *Hertfordshire Past and Present* 12, 3–9.

Austin, P. (2013) 'Pollards in early-modern south-east Hertfordshire', *The Local Historian* 43, 138–58.

Aveling, J.V. (1870) *The History of Roche Abbey*. Worksop.

Bailey, B. (ed.) (1996) *Northamptonshire in the Eighteenth Century: the drawings of Peter Tillemans and others*. Northampton: Northamptonshire Record Society Volume 39.

Bailey, J. (1810) *General View of the Agriculture of the County of Durham*. London.

Bailey, J. and Culley, G. (1794) *General View of the Agriculture of the County of Northumberland*. London.

Baird, W. and Tarrant, J. (1970) *Hedgerow Destruction in Norfolk 1946–1970*. Norwich: Centre of East Anglian Studies.

Baker, C., Moxey, P. and Oxford, P. (1978) 'Woodland continuity and change in Epping Forest', *Field Studies* 4, 645–69.

Barnes, G. and Williamson, T. (2006) *Hedgerow History: ecology, history and landscape character*. Macclesfield: Windgather Press.

Barnes, G. and Williamson, T. (2011) *Ancient Trees in the Landscape: Norfolk's arboreal heritage*. Oxford: Windgather.

Barnes, G. and Williamson, T. (2015) *Rethinking Ancient Woodland: the archaeology and history of woods in Norfolk*. Hatfield: University of Hertfordshire Press.

Barnes, G., Dallas, P. and Williamson, T. (2009) 'The black poplar in Norfolk', *Quarterly Journal of Forestry* 103, 31–8.

Barnes, P. (1984) 'The Economic History of Landed Estates in Norfolk since 1880'. Unpublished PhD thesis, University of East Anglia.

Barrett-Lennard, T. (1921) 'Two hundred years of estate management at Horsford during the 17th and 18th centuries', *Norfolk Archaeology* 20, 57–139.

Beale, J. (1657) *Herefordshire Orchards, a pattern for all England*. London.

Beaver, S.H. (1943) *The Land of Britain, Part 58: Northamptonshire; Part 59, the Soke of Peterborough*. London: Geographical Publications.

Beckett, J.V. (1984) 'The pattern of landownership in England and Wales 1660–1880', *Economic History Review* 38, 1–22.

Beckett, J.V. (1986) *The Aristocracy in England 1660–1914*. Oxford: Wiley-Blackwell.

Beckett, J.V. (1990) *The Agricultural Revolution*. Oxford: Wiley-Blackwell.

Bennett, K.D. (1983) 'Devensian late glacial and Flandrian vegetational history at Hockham Mere, Norfolk, England 1. Pollen percentages and concentrations', *New Phytologist* 95, 457–87.

Bennett, K.D. (1986) 'Competitive interactions among forest tree populations in Norfolk, England, during the last 10,000 years', *New Phytologist* 103, 603–20.

Bennett, K.D. (1988) 'Holocene pollen stratigraphy of central East Anglia, England', *New Phytologist* 109, 237–53.

Billingsley, J. (1794) *General View of the Agriculture of the County of Somerset*. London.

Bird, G. (1935) 'The ornithological life in Suffolk', *Proceedings of the Suffolk Naturalists Society* 3, xciv.

Birtles, S. (2003) '"A Green Space Beyond Self Interest": the evolution of common land in Norfolk'. Unpublished PhD thesis, University of East Anglia.

Blagrave, J. (1675) *The Epitome of the Art of Husbandry*. London.

Blith, W. (1649) *The English Improver Improve*. London.

Blome, R. (1686) *The Gentleman's Recreation*. London.

Blomefield, F. (1805) *An Essay towards a Topographical History of the County of Norfolk*, Vol. I. London.

Blomley, N. (2007) 'Making private property: enclosure, common right and the work of hedges', *Rural History* 18 (1), 1–21.

Boardman, E.T. (1939) 'The development of a Broadland estate at How Hill Ludham, Norfolk', *Transactions of the Norfolk and Norwich Naturalists' Society* 15, 14.

Bogart, D. (2005) 'Did turnpike trusts increase transportation investment in eighteenth-century England?' *The Journal of Economic History* 65 (2), 439–68.

Boutcher, W. (1775) *A Treatise on Forest-Trees*. Edinburgh.

Boys, J. (1805) *General View of the Agriculture of the County of Kent*. London.

Branch Johnson, W. (1970) *The Industrial Archaeology of Hertfordshire*. Newton Abbot: David and Charles.

Branch Johnson, W. (ed.) (1973) *Memorandums for … The Diary between 1798 and 1810 of John Carrington*. London and Chichester: Phillimore.

Brasier, C.M. (1991) 'Ophiostoma novo-ulmi sp. nov., causative agent of current Dutch elm disease pandemics', *Mycopathologia* 115, 151–61.

Brasier, C.M. (2008) 'The biosecurity threat to the UK and global environment from international trade in plants', *Plant Pathology* 57, 792–808.

Brasier, C.M. and Gibbs, J.M. (1973) 'Origin of the Dutch elm disease epidemic in Britain', *Nature* 242, 607–9.

Brenchley, W.E. and Adam, H. (1915) 'Recolonization of cultivated land allowed to revert to natural conditions', *Journal of Ecology* 3, 193–210.

Brown, A.F.J. (1996) *Prosperity and Poverty: rural Essex, 1700–1815*. Chelmsford: Essex Record Office Publications.

Brown, D. and Williamson, T. (2016) *Lancelot Brown and the Capability Men: landscape revolution in eighteenth-century England*. London: Reaktion Books.

Brown, J. (1847) *The Forester*. London.

Brown, J. (1871) *The Forester* (revised and extended edition). London.

Brown, N. and Fisher, R. (2009) *Trees Outside Woods: a Report to the Woodland Trust*. Grantham: The Woodland Trust.

Brown, N., Daegan, J.G. Inward, Jegar, M. and Denman, S. (2014) 'A review of *Agrilus biguttatus* in UK Forests and its relationship with Acute Oak Decline', *Forestry* 88 (1), 53–63.

Brown, R. (1799) *General View of the Agriculture of the West Riding of Yorkshire*. London.

Brown, T. and Foard, G. (1998) 'The Saxon landscape: a regional perspective'. In P. Everson and T. Williamson (eds) *The Archaeology of Landscape*. Manchester: Manchester University Press, 67–94.

Butcher, R.W. (1941) *The Land of Britain: Suffolk*. London: Geographical Publications.

Cameron, L.G. (1941) *The Land of Britain: Hertfordshire*. London: Geographical Publications.

Cameron, R.A.D. and Pannett, D.J. (1980) 'Hedgerow shrubs and landscape history in the West Midlands', *Arboricultural Journal* 4, 147–52.

Campbell, B.M.S. (1981) 'Commonfields origins: the regional dimension'. In T. Rowley (ed.) *The Origins of Open Field Agriculture*. London: Croom Helm, 112–29.

Campbell, G. and Robinson, M. (2008) 'Environment and land use in the valley bottom'. In J. Harding and F. Healy (eds) *The Raunds Area Project: a Neolithic and Bronze Age landscape in Northamptonshire*. Swindon, 19–28.

Carlton, J. (2012) *Marine Propellers and Propulsion*. London: Butterworth-Heinemann.

Chard, R. (1959) 'The Thetford Fire Plan', *Journal of the Forestry Commission* 28, 154–89.

Cheeseman, C. (2005) 'Ownership and ecological change'. In J. Langton and G. Jones (eds), *Forests and Chases of England and Wales c.1500–c.1800: towards a survey and analysis*. Oxford: St John's College/Oxbow Books, 69–74.

Cheffings, R. and Lawrence, C.M. (2014) *Chalara: a summary of the impacts of ash dieback on UK* biodiversity, including the potential for long term monitoring and future research on management scenarios. Peterborough: JNCC report No 501.

Chester, K.R. (1930) 'The phytopthora disease of the Culla in America', *Quarterly Journal of Forestry* 11, 169–71.

Cinotti, B. (1989) 'Winter moisture content and frost-crack occurrence in oak trees (*Quercus petraea Liebl. and Q. robur L.*)', *Annales des Sciences Forestierès* 46, 614–16.

Clarke, J. (1794) *General View of the Agriculture of the County of Hereford*. London.

Clarke, W.G. (1918) 'The natural history of Norfolk commons', *Transactions of the Norfolk and Norwich Naturalists' Society* 10, 294–318.

Clarkson, L.A. (1974) 'The English bark trade, 1660–1830', *Agricultural History Review* 22, 136–52.

Clemenson, H. (1982) *English Country Houses and Landed Estates*. London: Croom Helm.

Clouston, B. and Stansfield, K. (1979) *After the Elm*. London: Heinemann.

Cobbett, W. (1825) *The Woodlands*. London.

Colebourn, P.H. (1989) 'Discovering ancient woods', *British Wildlife* 1, 61–75.

Colley, L. (2009) *Britons: forging the nation 1707– 1837*. London: Yale University Press.

Collins, E.J.T. (1989) 'The coppice and underwood trades'. In G.E. Mingay (ed.) *The Agrarian History of England and Wales Vol. 6, 1750–1850*. Cambridge: Cambridge University Press, 484–501.

Cook, M. (1676) *On the Manner of Raising, Ordering and Improving Forest-Trees*. London.

Cooper, F. (2006) *The Black Poplar: history, ecology and conservation*. Oxford: Windgather.

Cossons, N. (1987) *The BP Book of Industrial Archaeology*. Newton Abbot: David and Charles.

Cowell, B. (2002) 'The commons preservation society and the campaign for Berkhamsted common, 1866–70', *Rural History* 13 (2), 145–62.

Cox, A. (1979) A *Survey of Bedfordshire Brickmaking: a history and gazetteer*. Bedford: RCHME and Bedfordshire County Council.

Crick, J. (ed.) (2007) *The Charters of St Albans*. Oxford: Oxford University Press.

Crosby, A. and Winchester, A. (2006) 'A sort of national property?' In A. Winchester (ed.), *England's Landscape: the North West*. London: Collins/English Heritage, 235–7.

Crowther, J. (1992) *Descriptions of East Yorkshire: De La Pryme to Head*. York: East Yorkshire Local History Society.

Culley, G. (1794) *General View of the Agriculture of the County of Cumberland*. London.

Curtis, C.E. (1892) *The Manifestation of Disease in Forest Trees: The causes and remedies*. London.

Dallas, P. (2010) 'Sustainable environments: common wood-pastures in Norfolk', *Landscape History* 31 (1), 23–36.

Daniels, S. (1988) 'The political iconography of woodland in later eighteenth-century England'. In D. Cosgrove and S. Daniels (eds), *The Iconography of Landscape*. Cambridge: Cambridge University Press, 51–72.

Dannatt, N. (1996) 'Thetford Forest: its history and development'. In P. Ratcliffe and J. Claridge (eds), *Thetford Forest Park: the ecology of a pine forest*. Edinburgh: Forestry Commission, 21–5.

Dark, P. (2000) *The Environment of Britain in the First Millennium AD*. London: Duckworth.

Davenport, F.G. (1906) *The Economic Development of a Norfolk Manor, 1086–1565*. Cambridge: Cambridge university Press.

Davies, R. (1964) *The Trade and Shipping of Hull 1500–1700*. York: East Yorkshire Local History Society.

Davis, B.N.K. (1976) 'Wildlife, urbanisation and industry', *Biological Conservation* 10, 249–91.

Davis, T. (1794) *General View of the Agriculture of the County of Wiltshire*. London.

Day, A. (1999) *Fuel from the Fens: the Fenland turf industry*. Cambridge: SB Publications.

Day, W.R. (1948) 'The penetration of conifer roots by *Fomes annosus*', *Quarterly Journal of Forestry* 42, 99–101.

Defoe, D. (1724) *A Tour Through This Whole Isalnd of Great Britain*, Vol. 1. London.

Denman, S. and Webber, J.F. (2009) 'Oak declines – new definitions and new episodes in Britain', *Quarterly Journal of Forestry* 103 (4), 285–90.

Denman, S., Brady, C., Kirk, S., Cleenwerck, I., Venter, S., Coutinho, T.A. and De Vos, P. (2012) '*Brenneria goodwinii sp. nov.*', associated with acute oak decline in the UK', *International Journal of Systematic and Evolutionary Microbiology* 62, 2451–6.

Denman, S., Brown, N., Kirk, S.A., Jeger, M. and Webber, J.F. (2014) 'A description of the symptoms of acute oak decline in Britain and a comparative review on causes of similar decline diseases of oak in Europe', *Forestry* 10, 1–17.

Dolman, P. and Waeber, W. (2008) 'Ecosystem and competition impacts of introduced deer', *Wildlife Research* 35 (3), 202–14.

Donaldson, J. (1794) *General View of the Agriculture of the County of Northampton*. London.

Donovan, E. (1799) *The Natural History of British Insects*, Vol 8. London.

Dormor, I. (2002) 'Woodland Management in Two Yorkshire Dales Since the Fifteenth Century'. Unpublished PhD thesis, University of Leeds.

Duncumb, J.M. (1805) *General View of the Agriculture of the County of Hereford*. London.

Edelen, G. (ed.) (1994) *The Description of England: the classic contemporary account of Tudor social life, by William Harrison*. London: Folger Shakespeare Library.

Eden, F.M. (1797) *The State of the Poor: or, an history of the labouring classes in England from the Conquest to the present period*. London.

Edlin, H.C. (1944) *British Woodland Trees*. London: Batsford.

Edlin, H.L. (1949) *Woodland Crafts in Britain*. London: Batsford.

Edlin, H.L. (1958) *England's Forests*. London: Faber and Faber.

Ehrlich, J. (1932) 'The occurrence in the United States of *Cryptococcus fagi* (Baer) Dougl. The insect factor in a menacing disease of beech', *Quarterly Journal of Forestry* 13, 73–80.

Ellis, W. (1741) *The Timber Tree Improv'd: or, the best practical methods of improving different lands with proper timber*. London.

Evans, D. (1992) *A History of Nature Conservation in Britain*. London: Routledge.

Evelyn, J. (1664) *Sylva, or a Discourse of Forest-Trees*. London.

Faden, W. (1797) *A Topographical Map of the County of Norfolk*. London.

Fischer, J., Stott, J. and Law, B.S. (2010) 'The disproportionate value of scattered trees', *Biological Conservation* 143, 1564–7.

Fisher, W.R. (1895) *Forest Protection*, forming volume 4 of W. Schlich, *Manual of Forestry*. 4 vols, London.

Fitzgerald, R. and Grenier, J. (1992) *A History of the Timber Trade Federation*. London: Batsford.

Fitzherbert, J. (1533) *The Boke of Husbandry*. London.

Fleming, A. (1998) *Swaledale: the valley of the wild river*. Edinburgh: Edinburgh University Press.

Flinn, M.W. (1984) *The History of the British Coal Industry Vol. 2, 1700–1830*. Oxford: Oxford University Press.

Foreman, D. (2004) *Rewilding North America: a vision for conservation in the 21st century*. Washington D.C.: Island Press.

Forestry Commission (1955) *Report of the Committee on Hedgerow and Farm Timber*. London: HMSO.

Forestry Commission (2002) *National Inventory of Woodland and Trees*. Edinburgh: Forestry Commission.

Forestry Commission (2016) *Forestry Statistics 2016: a compendium of statistics about woodland, forestry and primary wood processing in the United Kingdom*. Edinburgh: Forestry Commission.

Forman, R.T. and Baudry, J. (1984) 'Hedgerow and hedgerow networks in landscape ecology', *Environmental Management* 8, 495–510.

Fowler, P.J. (1983) *The Farming of Prehistoric Britain*. Cambridge: Cambridge University Press.

Fuller, R., Williamson, T., Barnes, G. and Dolman, P. (2017) 'Human activities and biodiversity opportunities in pre-industrial cultural landscapes: relevance to conservation', *Journal of Applied Ecology* 54, 459–69.

Furst, H. (1893) *The Protection of Woodlands Against Dangers Arising from Organic and Inorganic Causes*. Edinburgh.

George, M. (1992) *The Land Use, Ecology and Conservation of Broadland*. Chichester: Packard Publishing.

Gibbs, J.N., Brasier, C.M. and Webber, J.F. (1994) *The Biology of Dutch Elm Disease*. Edinburgh: Forestry Commission, Research Information Note. 252.

Gibbs, J., Greig, B. and Rishbeth, J. (1996) 'Tree diseases of Thetford Forest and their influence on its ecology and management'. In P. Ratcliffe and J. Claridge (eds), *Thetford Forest Park: the ecology of a pine forest*. Edinburgh: Forestry Commission, 26–32.

Gledhill, T. (1994) 'A Woodland History of North Yorkshire'. Unpublished PhD thesis, University of Sheffield.

Godwin, H. (1968) 'Studies in the post-glacial history of British vegetation 15. Organic deposits of Old Buckenham Mere, Norfolk', *New Phytologist* 67, 95–107.

Gooch, W. (1811) *General View of the Agriculture of the County of Cambridge*. London.

Gowans, E. and Pouncett, J. (2003) *Rapid Assessment of Q-Pits, Ecclesall Wood, Sheffield: archaeological survey*. Oxford: Archaeological Survey and Evaluation Ltd.

Granger, J. (1794) *General View of the Agriculture of the County of Durham*. London.

Green, T. (2005) 'Is there a case for the Celtic Maple or the Scots Plane?' *British Wildlife* 16, 184–8.

Greig, J.R.A. (1989) 'From lime forest to heathland – five thousand years of change at West Heath Spa, Hampstead'. In D. Collins and D. Lorrimer,

Excavations at the Mesolithic Site on West Heath, Hampstead 1976–1981. Oxford: British Archaeological Reports, 217, 89–99.

Grigor, J. (1841) *The Eastern Arboretum: or a register of remarkable trees, seats, gardens &c in the County of Norfolk*. London.

Hadfield, M. (1967) *Landscape with Trees*. London: Country Life.

Haggard, H.R. (1902) *Rural England*. London: Longmans.

Haggard, L.R. and Williamson, H. (1943) *Norfolk Life*. London: Faber.

Haggard, L.R. (1946) *Norfolk Notebook*. London: Faber.

Hainsworth, D.R. and Walker, C. (1990) *The Correspondence of Lord Fitzwilliam of Milton and Francis Guybon his Steward, 1697–1709*. Northampton: Northamptonshire Record Society 36.

Hale, T. (1756) *A Compleat Body of Husbandry*. London.

Hall, D. (1982) *Medieval Fields*. Princes Risborough: Shire Books.

Hall, D. (1995) *The Open Fields of Northamptonshire*. Northampton: Northamptonshire Record Society.

Hall, J. (1982) 'Hedgerows in west Yorkshire: the Hooper method examined', *Yorkshire Archaeological Journal* 54, 103–9.

Hamilton, G.J. (1820) 'An essay on woods and plantations', *Prize Essays and Transactions of the Highland Society of Scotland*. Edinburgh and London, 189–490.

Hammersley, G. (1973) 'The charcoal iron industry and its fuel', *Economic History Review* 26 (4), 593–613.

Hanbury, W. (1758) *An Essay on Planting*. London.

Hart, C. (1966) *Royal Forest: a History of Dean's Woods as Producers of Timber*. Oxford: Oxford University Press.

Hartig, R.H. (1894) *Textbook of Diseases of Trees*. London: MacMillan.

Hartlib, S. (1651) *An Enlargement of the Discourse of Husbandry*. London.

Hatcher, J. (1993) *The History of the British Coal Industry. Volume 1. Before 1700: towards the age of coal*. Oxford: Oxford University Press.

Hatley, V.A. (1980) 'Locks, lords and coal: a study in eighteenth-century Northamptonshire history', *Northamptonshire Past and Present* 6, 207–18.

Hill, D. and Robertson, P. (1988) *The Pheasant: ecology, management and conservation*. Oxford: BSP Professional.

Hodder, K.H., Buckland, P.C., Kirby, K.J. and Bullock, J.M. (2009) 'Can the pre-Neolithic provide suitable models for re-wilding the landscape in Britain?' *British Wildlife* 20 (5) (special supplement), 4–14.

Hodge, C.A.H., Burton, R.G.O., Corbett, W.M., Evans, R. and Searle, R.S. (1984) *Soils and their Use in Eastern England*. Harpenden: Soil Survey of England and Wales.

Holderness, B.A. (1989) 'Prices, productivity and output'. In G.E. Mingay (ed.), *The Agrarian History of England and Wales*, Volume VI. Cambridge: Cambridge University Press, 84–189.

Holland, H. (1808) *General View of the Agriculture of Cheshire*. London.

Holt, J. (1794) *General View of the Agriculture of the County of Lancaster*. London.

Holt, J.S. (1999) 'Roeburndale Woodlands', *Contrebis* 24, 14–20.

Howkins, C. and Sampson, N. (2000) *Searching for Hornbeam: a social history*. Addlestone: Chris Howkins.

Humphrey, J., Ferris, R. and Peace, A. (2003) *Biodiversity in Britain's Planted Forests*. Edinburgh: Forestry Commission.

Hunter, J. (1999) *The Essex Landscape: a study of its form and history*. Chelmsford: Essex Record Office.

Jackson, C. (ed.) (1870) *The Diary of Abraham de la Pryme, the Yorkshire Antiquary*. Durham: Publications of the Surtees Society 54.

Jackson, G. (1975) *The Trade and Shipping of Eighteenth Century Hull*. York: East Yorkshire Local History Society.

James, N.D.G. (1990) *A History of English Forestry*. Oxford: Blackwell.

Jefferys, T. (1771) *The County of York Survey'd in MDCCLXVII, VIII, IX and MDCCLXX*. London.

Jepson, P. (2015) 'A rewilding agenda for Europe: creating a network of experimental reserves', *Ecography* 36, 1–8.

Jessopp, A. (1887) 'Beeston Priory, otherwise Moulney', *Norfolk Antiquarians Miscellany* 3, 439–61.

Johns, C.A. (1899) *The Forest Trees of Britain*. London.

Johnson, W. (1978) 'Hedges: a review of some early literature', *Local Historian* 13, 195–204.

Jones, M. (1993) 'South Yorkshire's ancient woodland: the historical evidence'. In I. Rotherham and P. Beswick (eds), *Ancient Woodlands, their Archaeology and Ecology: a coincidence of interest*. Sheffield: Landscape Conservation Forum, 26–48.

Jones, M. (1997) 'Woodland management on the Duke of Norfolk's Sheffield estate in the early eighteenth century'. In M. Jones (ed.), *Aspects of Sheffield: discovering local history*. Trowbridge: Wharncliffe, 48–69.

Jones, M. (2005) 'Ancient woodland destruction, survival and restoration: a South Yorkshire perspective', *Landscape Archaeology and Ecology* 5, 46–58.

Jones, M. (2012) *Trees and Woodlands in the South Yorkshire Landscape*. Barnsley: Wharncliffe.

Journal of the House of Commons (1792).

Kent, N. (1796) *General View of the Agriculture of the County of Norfolk*. London.

Kerridge, E. (1967) *The Agricultural Revolution*. London: Routledge.

Kerridge, E. (1993) *The Common Fields of England*. Manchester: Manchester University Press.

King, H. and Harris, A. (eds) (1963) *A Survey of the Manor of Settrington*. Leeds: Yorkshire Archaeological Society Record Series Vol. 126.

Kirby, J. and Baker, A. (2013) 'The dynamics of pre-Neolithic European landscapes and their relevance to modern conservation'. In I.D. Rotherham (ed.), *Trees, Forested Landscapes and Grazing Animals*. London: Routledge, 87–98.

Langley, B. (1728) *A Sure Method of Improving Estates*. London.

Langton, J. (1979) *Geographical Change and Industrial Revolution. Coal mining in south-west Lancashire 1590-1799*. Cambridge: Cambridge University Press.

Langton, J. (2005) 'Forests in early modern England and Wales: history and historiography'. In J. Langton and G. Jones (eds), *Forests and Chases of England and Wales c.1500–c.1800: towards a survey and analysis*. Oxford: St John's College/Oxbow Books, 1–9.

Latham, B. (1957) *Timber. Its Development and Distribution: an historical survey*. London: George Harrap and Co.

Lawes, J.B. (1895) *The Rothamsted Experiments*. London.

Lawson, W. (1618) *A New Orchard and Garden*. London.

Lennon, B. (2009) 'Estimating the age of groups of trees in historic landscapes', *Arboricultural Journal* 32, 167–88.

Linnell, J.E. (1932) *Old Oak: the story of a forest village*. London: Constable and Co.

Loudon, J.C. (1838) *Arboretum et Fruticetum Britannicum; or, the trees and shrubs of Britain*. London.

Lowe, R. (1794) *General View of the Agriculture of the County of Nottingham*. London.

Lucas, J. (ed.) (1892) *Kalm's Account of his Visit to England on his Way to America in 1748*. London.

Lyons, S.K., Smith, F.A. and Brown, J.H. (2004) 'Of mice, mastodons and men: human-mediated extinctions on four continents', *Evolutionary Ecology Research* 6, 339–58.

MacCulloch, D. (2007) *Letters from Redgrave Hall: The Bacon Family, 1340–1744*. Woodbridge: Boydell for the Suffolk Records Society.

McDonagh, B.A.K. (2009) 'Subverting the ground: private property and public protest in the sixteenth-century Yorkshire Wolds', *Agricultural History Review* 57 (2), 191–206.

MacNair, A. and Williamson, T. (2010) *William Faden and the Norfolk Eighteenth-Century Landscape*. Oxford: Windgather.

MacNair, A.D.M, Rowe, A. and Williamson, T. (2015) *Dury and Andrews' Map of Hertfordshire: society and landscape in the eighteenth century*. Oxford: Windgather Press.

Marcais, B. and Desprez-Loustau, M. (2014) 'European oak powdery mildew: impact on trees, effects of environmental factors, and potential effects of climate change', *Annals of Forest Science* 71, 633–42.

Marshall, W. (1787) *The Rural Economy of Norfolk*, 2 volumes. London.

Marshall, W. (1789) *The Rural Economy of Gloucestershire*. London.

Marshall, W. (1796) *Planting and Rural Ornament*. London.

Martin, E. and Satchell, M. (2008) *Wheare Most Inclosures Be. East Anglian Fields, History, Morphology and Management*. Ipswich: East Anglian Archaeology vol. 124.

Mather, R.A. and Saville, P.S. (1994) 'The commercial impact of oak shake in Great Britain', *Forestry* 67 (2), 119–31.

Maw, P.T. (1910) 'The importation of foreign grown seedlings', *Quarterly Journal of Forestry* 4, 267–8.

Maw, P.T. (1912) *The Practice of Forestry*. London: Fisher Unwin.

Mawer, A. and Stenton, F.M. (1938) *The Place-Names of Hertfordshire*. Cambridge: Cambridge University Press.

Meager, L. (1697) *The Mystery of Husbandry: or, arable pasture and wood-land improved*. London.

Miche, C.Y. (1888) *The Practice of Forestry*. London.

Middleton, J. (1798) *View of the Agriculture of Middlesex*. London.

Miller, P. (1731) *The Gardener's Dictionary*. London.

Mitchell, B.R. and Deane, P. (1962) *Abstract of British Historical Statistics*. Cambridge: Cambridge University Press.

Monbiot, G. (2015) *Feral: searching for enchantment on the frontiers of rewilding*. London: Penguin.

Monk, J. (1794) *General View of the Agriculture of the County of Leicester*. London.

Monteath, R. (1824) *The Forester's Guide and Profitable Planter*. London.

Moraal, L.G. and Hilszczanski, J. (2000) 'The oak buprestid beetle, *Agrilus biguttatus* (F.) (Col. Buprestidae), a recent factor in oak decline in Europe', *Journal of Pest Science* 73 (5), 134–8.

Moreton, C. and Rutledge, P. (eds) (1997) 'Skayman's Book. 1516–1518'. In C. Noble, C. Moreton and P. Rutledge (eds), *Farming and Gardening in Late Medieval Norfolk*. Norwich: Norfolk Record Society, 95–155.

Mortimer, J. (1707) *The Whole Art of Husbandry: or, the way of managing and improving of land*. London.

Morton, J. (1712) *The Natural History of Northamptonshire*. London.

Mosby, J.E.G. (1938) *The Land of Britain: Norfolk*. London: Geographical Publications.

Mougou, A., Dutech, C. and Desprez-Loustau, M.-L. (2014) 'New insights into the identity and origin of the causal agent of oak powdery mildew in Europe', *Forest Pathology* 38 (4), 275–87.

Muir, R. (2000) 'Pollards in Nidderdale: a landscape history', *Rural History: economy, society and culture* 11 (1), 95–111.

Munby, L. (1977) *The Hertfordshire Landscape*. London: Hodder and Stoughton.

Murray, A. (1813) *General View of the Agriculture of the County of Warwick*. London.

Nichols, J. (1795) *The History and Antiquities of the County of Leicester*. London.

Nicol, W. (1799) *The Practical Planter, or, a Treatise on Forest Planting*. London.

Nicol, W. (1820) *The Planters Kalendar*. Edinburgh.

Nisbet, J. (1894) *John Brown, The Forester: enlarged and revised in 2 vols*. London.

Norden, J. (1608) *The Surveyor's Dialogue*. London.

Nourse, T. (1699) *Campania Felix: or, a discourse of the benefits and improvements of husbandry*. London.

Orchard, N. (2007) 'Further Changes in Rockingham Forest Woodlands'. Unpublished MSc dissertation, Birkbeck College, London.

Overton, M. (1996) *Agricultural Revolution in England: the transformation of the agrarian economy, 1500–1850*. Cambridge: Cambridge University Press.

Owen, J. (2010) *Wildlife of a Garden: a thirty year study*. Peterborough: Royal Horticultural Society.

Page, W. (ed.) (1902) *The Victoria History of the County of Hertford*, Vol. 1. London: Archibald Constable.

Page, W. (ed.) (1908) *The Victoria History of the County of Hertford*, Vol. 2. London: Archibald Constable.

Page, W. (ed.) (1912) *The Victoria History of the County of Hertford*, Vol. 3. London: Archibald Constable.

Paley, R. (2005) 'Parliament, peers and legislation, 1660–1900'. In J. Langton and G. Jones (eds), *Forests and Chases of England and Wales c.1500–c.1800: towards a survey and analysis*. Oxford: St John's College/Oxbow Books, 29–32.

Palmer, M. and Neaverson, P. (1994) *Industry in the Landscape 1700–1900*. London: Routledge.

Parkinson, R. (1808) *General View of the Agriculture of the County of Rutland*. London.

Parsons, M.A. (1997) 'The woodland of Troutbeck and its exploitation to 1800', *Transactions of the Cumberland and Westmoreland Antiquarian and Archaeological Society* 97, 79–100.

Partida, T. (2014) 'Drawing the Lines: a GIS study of enclosure and landscape in Northamptonshire'. Unpublished PhD thesis, University of Huddersfield.

Pearce, W. (1794) *General View of the Agriculture in Berkshire*. London.

Peglar, S., Fritz, S. and Birks, H. (1989) 'Vegetation and land use history at Diss, Norfolk', *Journal of Ecology* 77, 203–22.

Pennant, T. (1776) *A Tour in Scotland, and Voyage to the Hebrides; MDCCLXXII, Part I*. London.

Perren, R. (1995) *Agriculture in Depression 1870–1940*. Cambridge: Cambridge University Press.

Perry, P.J. (1974) *British Farming in the Great Depression*. Newton Abbot: David and Charles.

Peterken, G.F. (1976) 'Long-term changes in the woodlands of Rockingham Forest and other areas', *Journal of Ecology* 64 (1), 123–46.

Peterken, G.F. (1981) *Woodland Conservation and Management*. Cambridge: Cambridge University Press.

Peterken, G.F. (1996) *Natural Woodland: ecology and conservation in northern temperate regions*. Cambridge: Cambridge University Press.

Peterken, G.F. and Allison, H. (1989) *Woods, Trees and Hedges: a review of changes in the British countryside. Focus on nature conservation*. Peterborough: Nature Conservancy Council.

Petit, S. and Watkins, C. (2003) 'Pollarding trees; changing attitudes to a traditional land management practice in Britain, 1600–1900', *Rural History* 14, 157–76.

Pettit, P.A. (1968) *The Royal Forests of Northamptonshire: a study of their economy 1558–1714*. Northampton: Northamptonshire Record Society.

Phillips, R. (1810) *General View of the Agriculture of Hampshire, including the Isle of Wight*. London.

Pigot, D. (2012) *Lime-trees and Basswoods. A biographical monograph on the genus Tilia*. Cambridge.

Pitt, W. (1796) *General View of the Agriculture of the County of Stafford*. London.

Pitt, W. (1810) *General View of the Agriculture of the County of Worcester*. London.

Plot, R. (1686) *The Natural History of Staffordshire*. Oxford.

Plymley, J. (1803) *General View of the Agriculture of Shropshire*. London.

Pollard, E., Hooper, M.D. and Moore, N.W. (1974) *Hedges*. London: Collins.

Pontey, W. (1805) *The Forest Pruner, or Timber-owner's Assistant*. London.

Price, A. (2015) *Shake in Oak: an evidence review*. Edinburgh: Forestry Commission Research Report.

Prince, H. (1987) 'The changing rural landscape'. In G. Mingay (ed.), *The Cambridge Agrarian History of England and Wales*, Vol. 6. Cambridge: Cambridge University Press, 7–83.

Pringle, A. (1794) *General View of the Agriculture of the County of Westmoreland*. London.

Pryor, F. (1998) *Farmers in Prehistoric Britain*. Stroud: Tempus.

Rackham, O. (1976) *Trees and Woodlands in the British Landscape*. London: Dent.

Rackham, O. (1980) *Ancient Woodland*. London: Edward Arnold.

Rackham, O. (1986) *The History of the Countryside*. London: Dent.

Rackham, O. (1989) *The Last Forest: the story of Hatfield Forest*. London: Dent.

Rackham, O. (2004) 'Pre-existing trees and woods in country-house parks', *Landscapes* 5, 1–15.

Rackham, O. (2006) *Woodlands*. London: Collins.

Raynbird, W. and Raynbird, H. (1849) *On the Farming of Suffolk*. London.

Read, H.J. (2008) 'Pollards and pollarding in Europe', *British Wildlife* 19, 250–59.

Reed, M. (1981) 'Pre-parliamentary enclosure in the east Midlands, 1550 to 1750, and its impact on the landscape', *Landscape History* 3, 60–68.

Rishbeth, J. (1963) 'Stump protection against *Fomes annosus*: inoculation with *Peniophora gigantea*', *Annals of Applied Biology* 52, 63–77.

Roberts, B.K. and Wrathmell, S. (2002) *Region and Place: a study of English rural settlement*. London: English Heritage.

Roden, D. (1973) 'Field systems of the Chiltern Hills and their environs'. In A.H.R. Baker and R.A. Butlin (eds), *Studies of Field Systems in the British Isles*. Cambridge: Cambridge University Press, 325–74.

Rollinson, W. (1974) *Life and Tradition in the Lake District*. Worthing: Littlehampton Books.

Ross, J.M. (1935) 'Study of Pine Shoot Moth damage', *Journal of the Forestry Commission* 14, 56–64.

Rotherham, I.D. (2011) *Peat and Peat Cutting*. Princes Risborough: Shire.

Rotherham, I.D. (2012), 'Searching for shadows and ghosts'. In I.D. Rotherham, C. Handley, M. Agnoletti and T. Somojlik (eds), *Trees Beyond the Wood: an exploration of concepts of woods, forests and trees*. Sheffield: Wildtrack Publishing, 1–16.

Rotherham, I.D., Jones, M., Smith, L. and Handley, C. (eds) (2008) *The Woodland Heritage Manual: a guide to investigating woodland landscapes*. Sheffield: Wildtrack Publishing.

Rowe, A. (2009) *Medieval Parks of Hertfordshire*. Hatfield: University of Hertfordshire Press.

Rowe, A. (2015) 'Pollards: living archaeology'. In K. Lockyear (ed.), *Archaeology in Hertfordshire: recent research*. Hatfield: University of Hertfordshire Press, 302–24.

Rowe, A. and Williamson, T. (2013) *Hertfordshire: a landscape history*. Hatfield: University of Hertfordshire Press.

Rowley, T. (2006) *The English Landscape in the Twentieth Century*. London: Hambledon Continuum.

Rudge, T. (1807) *General View of the Agriculture of the County of Gloucester*. London.

Ruggles, T. (1786) 'Picturesque farming', *Annals of Agriculture* 6, 175–84.

Ryden, L., Pawel, M. and Anderson, M. (2003) *Environmental Science: understanding, protecting and managing the environment of the Baltic Sea region*. Uppsala: Baltic University Press.

Ryle, G. (1969) *Forest Service: the first forty-five years of the Forestry Commission in Great Britain*. Newton Abbot: David and Charles.

Samojlik, T. and Kuijper, D. (2013) 'Grazed wood pasture versus browsed high forests: impact of ungulates on forest landscapes from the perspective of the Białowieża primeval forest'. In I.D. Rotherham (ed.), *Trees, Forested Landscapes and Grazing Animals: a European perspective on woodlands and grazed treescapes*. London: Routledge, 143–61.

Scarfe, N.V. (1942) *The Land of Britain: Essex*. London: Geographical Publications.

Selby, J. (1842) *A History of British Forest Trees*. London.

Sheail, J. (2002) *An Environmental History of Twentieth-Century Britain*. London: Palgrave.

Simmons, I.G. (2003) *The Moorlands of England and Wales: an environmental history 8000 BC to 2000 AD*. Edinburgh: Edinburgh University Press.

Simpson, J. (1900) *The New Forestry: or, the Continental system adapted to British woodlands and game preservation*. Sheffield: Pawson and Brailsford.

Skipper, K. and Williamson, T. (1997) *Thetford Forest: making a landscape, 1922–1997*. Norwich: Centre of East Anglian Studies.

Smith, A.H., Baker, G.M. and Kenny, R.W. (1982) *The Papers of Nathaniel Bacon of Stiffkey*, Vol. 2. Norwich: Centre of East Anglian Studies.

Smith, J. (1670) *England's Improvement Reviv'd*. London.

Soulé, M. and Noss, R. (1998) 'Rewilding and biodiversity: complementary goals for continental conservation', *Wild Earth* 8, 19–28.

Spencer, J. and Kirby, K. (1992) 'An inventory of ancient woodland for England and Wales', *Biological Conservation* 62, 77–93.

Spray, M. (1981) 'Holly as fodder in England', *Agricultural History Review* 29 (2), 97–110.

Stevenson, W. (1809) *General View of the Agriculture of the County of Surrey*. London.

Stone, A. and Williamson, T. (2013) '"Pseudo-ancient woodland" and the *Ancient Woodland Inventory*', *Landscapes* 14, 141–54.

Straker, V., Brown, A., Fyfe, R. and Jones, J. (2007) 'Romano British environmental background'. In C.J. Webster (ed.), *The Archaeology of South-West England*. Taunton: Somerset Heritage Service, 145–50.

Straw, N.A. and Williams, D.T. (2013) 'Impact of the leaf miner *Cameraria ohridella* (Lepidoptera: Gracillariidae) and bleeding canker disease on horse-chestnut: direct effects and interaction', *Agricultural and Forest Entomology* 15 (3), 321–33.

Switzer, S. (1718) *Ichnographica Rustica*. London.

Szabo, P. (2013) 'Re-thinking pannage and historical interactions'. In I.D. Rotherham (ed.), *Trees, Forested Landscapes and Grazing Animals: a European perspective on woodlands and grazed treescapes*. London: Routledge, 51–60.

Tabbush, P. and Beaton, A. (1998) 'Hybrid poplars: present status and potential in Britain', *Forestry* 71 (4), 355–64.

Takamatsu, S., Braun, U., Limkaisang, S., Kom-Un, S., Sato, Y. and Cunnington, J.H. (2007) 'Phylogeny and taxonomy of the oak powdery mildew *Erysiphe alphitoides sensu lato*', *Mycological Research* 111, 809–26.

Tallis, J.H. (1991) 'Forest and moorland in the south Pennine uplands in the mid-Flandrian period III. The spread of moorland – local, regional and national', *Journal of Ecology* 79, 401–15.

Tansley, A.G. (1949) *The British Islands and their Vegetation*, Vol. 1. Cambridge: Cambridge University Press.

Taylor, C. (1975) *Fields in the English Landscape*. London: Dent.

Tennyson, J. (1939) *Suffolk Scene: a book of description and adventure*. London: Blackie.

Theobald, J. 'Changing Landscapes, Changing Economies: holdings in woodland High Suffolk 1600–1850'. Unpublished PhD thesis, University of East Anglia.

Thirsk, J. (1987) *England's Agricultural Regions and Agrarian History 1500–1750*. London: MacMillan.

Thomas, F.M., Blank, R. and Hartmann G. (2002) 'Abiotic and biotic factors and their interactions as causes of oak decline in Central Europe', *Forest Pathology* 32, 277–307.

Thomas, K. (1983) *Man and the Natural World: changing attitudes in England 1500–1800*. London: Allen Lane.

Thompson, F.M.L. (1963) *English Landed Society in the Nineteenth Century*. London: Routledge and Kegan Paul.

Thompson, F.M.L. (1968) 'The second Agricultural Revolution', *Economic History Review* 21, 62–77.

Tilbury, C. and Evans, H. (2003) *Exotic Pest Alert. Horse Chestnut leaf miner*. Edinburgh: Forestry Commission.

Toseland, A. (2013) *Estate Letters from the Time of the 2nd Duke of Montagu, 1709–32*. Northampton: Northamptonshire Record Society Vol. 46.

Toulmin-Smith, L. (ed.) (1907) *The itinerary of John Leland in or about the years 1535–1543*. London: George Bell.

Tuke, J. (1800) *General View of the Agriculture of the North Riding of Yorkshire*. London.

Turner, G. (1794) *General View of the Agriculture of the County of Gloucestershire*. London.

Turner, M.E. (1982) *Volume 190: Home Office Acreage Returns HO67. List and Analysis*, 3 vols. London: List and Index Society.

Tusser, T. (1573) *Five Hundred Good Points of Husbandrie*. London.

Vancouver, C. (1795) *General View of the Agriculture of the County of Essex*. London.

Vancouver, C. (1810) *General View of the Agriculture of Hampshire*. London.

Vera, F. (2002) *Grazing Ecology and Forest History*. Wallingford: Cabi.

Wade, E. (1755) *A Proposal for Improving and Adorning the Island of Great Britain; for the maintenance of our navy and shipping*. London.

Wade Martins, S. and Williamson, T. (eds) (1995) *The Farming Journal of Randall Burroughes of Wymondham, 1794–99*. Norwich: Norfolk Records Society.

Wade Martins, S. and Williamson, T. (1999) *Roots of Change: farming and the landscape in East Anglia, c.1700–1870*. Exeter: British Agricultural History Society.

Wake, J. and Webster, D.C. (1971) *The Letters of Daniel Eaton to the Third Earl of Cardigan, 1725–1732*. Northampton: Northamptonshire Record Society.

Walker, D. (1795) *General View of the Agriculture of the County of Hertford*. London.

Walsingham, Lord and Payne-Gallwey, R. (1983) *Shooting: field and covert*. London.

Ward, H. Marshall (1892) *The Oak: a popular introduction to forest-botany*. London.

Ward, J.R. (1967) *East Yorkshire Landed Estates in the Nineteenth Century*. York: East Yorkshire Local History Society.

Warde, P. (2006) *Energy Consumption in England and Wales 1560–2000*. Rome: Instituto di Studio sulle Societa del Mediterrane.

Warde, P. and Williamson, T. (2014) '*Fuel* supply and agriculture in post-medieval England', *Agricultural History Review* 62, 61–82.

Wargo, P.M. (1996) 'Consequences of environmental stress on oak: predisposition to pathogens', *Annals of Forest Science* 53, 339–68.

Warner, P. (1987) *Greens, Commons and Clayland Colonization*. Leicester: Leicester University Press.

Wass, J.G. (1956) '*Fomes Annosus* in East Anglian pine sample plots', *Journal of the Forestry Commission* 9, 75.

Waterson, M. (1994) *The National Trust: the first hundred years*. London: BBC Publications.

Watkins, C. (1990) *Britain's Ancient Woodland: woodland management and conservation*. Newton Abbot: David and Charles.

Watkins, C. (2014) *Trees, Woods and Forests: a social and cultural history*. London: Reaktion Books.

Watson, P.V. (1982) 'Man's impact on the chalklands: some new pollen evidence'. In M. Bell and S. Limbrey (eds), *Archaeological Aspects of Woodland Ecology*. Oxford: British Archaeological Reports International Series 146, 75–91.

West, T. (1774) *The Antiquities of Furness*. London.

Westmacott, R. and Worthington, T. (1974) *New Agricultural Landscapes*. London: Countryside Commission.

Wheeler, J. (1747) *The Modern Druid*. London.

White, J. (1998) 'Estimating the age of large and veteran trees in Britain'. Forestry Commission Information Note 250.

Whitelock, D. (1955) *English Historical Documents I: 500–1042*. London: Eyre and Spottiswoode.

Whynbrow, G.H. (1934) *The History of Berkhamsted Common*. London: Commons, Open Spaces and Footpaths Preservation Society.

Whyte, I. (2003) *Transforming Fell and Valley: Landscape and Parliamentary Enclosure in North West England*. Lancaster.

Williams-Ellis, C. (1928) *England and the Octopus*. London: G. Bles.

Williamson, T.M. (1998) *The Archaeology of the Landscape Park: garden design in Norfolk, England, c.1680–1840*. Oxford: British Archaeological Reports, British Series, 268.

Williamson, T. (2000) 'Understanding enclosure', *Landscapes* 1, 56–79.

Williamson, T. (2002) *The Transformation of Rural England: farming and the landscape 1700–1870*. Exeter: Exeter University Press.

Williamson, T. (2007) 'Archaeological perspectives on landed estates: research agendas'. In J. Finch and K. Giles (eds), *Estate landscapes: design, improvement and power in the post-medieval landscape*. Woodbridge: Boydell, 1–18.

Williamson, T. (2013a) *Environment, Society and Landscape in Early Medieval England: time and topography*. Woodbridge: Boydell.

Williamson, T. (2013b). *An Environmental History of Wildlife in England, 1550–1950*. London: Bloomsbury.

Williamson, T., Liddiard, R. and Partida, T. (2013) *Champion. The making and unmaking of the English Midland Landscape*. Liverpool: Liverpool University Press.

Willmott, A. (1980) 'The woody species of hedges with special reference to age in Church Broughton parish, Derbyshire', *Journal of Ecology* 68, 269–86.

Wilson, R. (2002) 'A journal of a tour through Suffolk, Norfolk, Lincolnshire and Yorkshire in the summer of 1741'. In C. Harper-Bill, C. Rawcliffe and R.G. Wilson (eds), *East Anglia's History: studies in honour of Norman Scarfe*. Woodbridge: Boydell.

Witney, K.P. (1998) 'The woodland economy of Kent, 1066–1348', *Agricultural History Review* 38, 20–39.

Woodward, D. (ed.) (1985) *Descriptions of East Yorkshire: From Leland to Defoe*. York: East Yorkshire Local History Society.

Wooldridge, S.N. (1945) *The Land of Britain, Part 51: Yorkshire, North Riding*. London: Geographical Publications.

Worlidge, J. (1660) *A Compleat System of Husbandry and Gardening*. London.

Wrigley, E.A. (2010) *Energy and the English Industrial Revolution*. Cambridge: Cambridge University Press.

Yalden, D.W. (1999) *The History of British Mammals*. London: Poyser.

Yalden, D. (2013) 'The post-glacial history of grazing animals in Europe'. In I.D. Rotherham (ed.), *Trees, Forested Landscapes and Grazing Animals: a European perspective on woodlands and grazed treescapes*. London: Routledge, 62–9.

Yelling, J.A. (1977) *Common Field and Enclosure in England 1450–1850*. London: Archon.

Young, A. (1797) *General View of the Agriculture of the County of Suffolk*. London.

Young, A. (1804) *General View of the Agriculture of Hertfordshire*. London.

Young, A. (1807) *General View of the Agriculture of the County of Essex*. London.

Young, A. (1808) *General View of the Agriculture of the County of Sussex*. London.

Young, A. (1809) *General View of the Agriculture of Oxfordshire*. London.

Parliamentary Papers

The Eleventh Report Of The Commissioners Appointed To Enquire Into The State and Condition Of The Woods, Forests, and Land Revenues Of The Crown, And To Sell Or Alienate Fee Farm and other Unimproveable Rents (6 February 1792).

Board of Agriculture Returns, 1896.

Board of Agriculture Statistics, 1910.

Board of Agriculture Statistics, 1914.

Forestry Commission 4th Annual Report, 1924.

Forestry Commission 15th Annual Report, 1935.

Forestry Commission Annual Report, 1945.

Index

Entries in bold refer to the Figures